Java程序设计实验指导

赖小平 主 编
邬卓恒 副主编

清华大学出版社
北京

内容简介

Java 是一种完全面向对象的程序设计语言,具有通用性、高效性、平台移植性和安全性,得到了广泛的应用。在全球云计算和移动互联网的环境下,Java 更具备显著的优势和广阔的前景。本书通过大量的案例解析和上机实验使读者掌握面向对象的思想和面向对象的程序设计方法。全书共 14 章,每章由知识提炼、实例解析、上机实验、拓展练习组成。在进行实验之前,先了解相关的知识点、实验目的,然后通过典型实例的学习完成相应的实验内容,最后学有余力的学生可以完成拓展练习,以达到巩固提高的目的。

本书概念清晰,结构合理,叙述简明易懂。无论是编程新手,还是具有编程基础的读者,都可从本书中获得新知识。本书可作为 Java 程序设计教学的辅助用书,也可作为实践、考研的参考用书。

本书封面贴有清华大学出版社防伪标签,无标签者不得销售。
版权所有,侵权必究。举报:010-62782989,beiqinquan@tup.tsinghua.edu.cn。

图书在版编目(CIP)数据

Java 程序设计实验指导/赖小平主编. —北京:清华大学出版社,2021.1
ISBN 978-7-302-54586-6

Ⅰ.①J… Ⅱ.①赖… Ⅲ.①JAVA 语言-程序设计-高等学校-教学参考资料 Ⅳ.①TP312.8

中国版本图书馆 CIP 数据核字(2019)第 296152 号

责任编辑:刘翰鹏
封面设计:傅瑞学
责任校对:袁 芳
责任印制:刘海龙

出版发行:清华大学出版社
网　　址:http://www.tup.com.cn,http://www.wqbook.com
地　　址:北京清华大学学研大厦 A 座
邮　　编:100084
社 总 机:010-62770175
邮　　购:010-62786544
投稿与读者服务:010-62776969,c-service@tup.tsinghua.edu.cn
质量反馈:010-62772015,zhiliang@tup.tsinghua.edu.cn

印 装 者:三河市吉祥印务有限公司
经　　销:全国新华书店
开　　本:185mm×260mm
印　　张:14
字　　数:337 千字
版　　次:2021 年 1 月第 1 版
印　　次:2021 年 1 月第 1 次印刷
定　　价:39.90 元

产品编号:084822-01

前言

"Java 程序设计"是计算机类专业的专业基础课程,是一门实践性很强的课程。本书可作为 Java 程序设计教学的辅助用书,也可作为实践、考研的参考用书。本书打破传统的单一辅导书的编写形式,以 Java SE 7.0 为基础,注重可读性和实用性,对 Java 程序设计基础知识进行了归纳和提炼,通过条理清晰的知识归纳和通俗易懂的案例学习,使学生提高和巩固所学的 Java 程序设计知识。

本书共分 14 章。第 1 章为 Java 入门,主要学习掌握 Java 开发环境与开发工具,了解如何编写一个简单的应用程序;第 2~6 章为 Java 基础语法部分,包括上机训练变量、数据类型、运算符、选择结构、循环结构、Java 方法的定义与调用、数组的应用等。第 7~9 章为面向对象程序设计,包括上机实验类和对象、继承与多态、接口与包;第 10 章为文件读/写,上机实验使用 I/O 流完成文件的读/写操作;第 11 章为泛型与集合,上机实验泛型以及常见的集合框架用法;第 12 章为 GUI 编程,上机实验 Java 的图形界面技术,能够编写简单的 GUI 应用程序;第 13 章为 JDBC 编程,上机实验数据库编程技术,能够编写简单的管理信息系统;第 14 章为网络编程,上机实验网络编程技术,实现简单的网络通信功能。

每章由知识提炼、实例解析、上机实验、拓展练习组成。在进行实验之前,先了解相关的知识点、实验目的,然后通过典型实例的学习完成相应的实验内容,最后学有余力的学生可完成拓展练习,以达到巩固提高的目的。本书的应用实例全部在 JDK 1.7 环境下编译通过。

本书由广东交通职业技术学院赖小平策划和统稿,并与广东理工学院邬卓恒、王晓丽和杨泽共同完成书稿的编写和审核工作,其中第 1~6 章由赖小平编写,第 7~10 章由邬卓恒编写,第 11、12 章由王晓丽编写,第 13、14 章由杨泽编写。

本书在编写过程中参考了一些国内外优秀教材和实验教程,在此表示衷心的感谢。由于编者水平有限,书中难免有不足之处,恳请读者批评、指正。

编　者
2020 年 9 月

目 录

第 1 章　Java 入门 ··· 001

  1.1　知识提炼 ·· 001

    1.1.1　Java 运行平台 ··· 001

    1.1.2　Java 程序的运行机制 ·· 001

    1.1.3　Java SE 开发环境 ··· 002

    1.1.4　Java 开发工具 ··· 003

    1.1.5　Java 应用程序开发过程 ·· 008

  1.2　实例解析 ·· 009

  1.3　上机实验 ·· 010

  1.4　拓展练习 ·· 012

第 2 章　Java 基础语法 ··· 014

  2.1　知识提炼 ·· 014

    2.1.1　变量与常量 ·· 014

    2.1.2　基本数据类型 ··· 015

    2.1.3　运算符 ·· 018

    2.1.4　语句与复合语句 ·· 020

  2.2　实例解析 ·· 021

  2.3　上机实验 ·· 022

  2.4　拓展练习 ·· 024

第 3 章　选择结构 ··· 026

  3.1　知识提炼 ·· 026

    3.1.1　if 选择结构 ··· 026

    3.1.2　if-else 选择结构 ·· 026

    3.1.3　多重 if 选择结构 ··· 027

    3.1.4　switch 结构 ·· 028

    3.1.5　if 与 switch 的比较 ·· 028

  3.2　实例解析 ·· 029

  3.3　上机实验 ·· 032

3.4 拓展练习 ……………………………………………………………………… 033

第 4 章 循环结构 …………………………………………………………………… 036

4.1 知识提炼 ………………………………………………………………………… 036
    4.1.1 while 语句 ……………………………………………………………… 036
    4.1.2 do-while 语句 …………………………………………………………… 037
    4.1.3 for 语句 ………………………………………………………………… 038
    4.1.4 break 与 continue 语句 ………………………………………………… 039
    4.1.5 多重循环 ………………………………………………………………… 039
4.2 实例解析 ………………………………………………………………………… 040
    4.2.1 实例 1：累加程序 ……………………………………………………… 040
    4.2.2 实例 2：乘法口诀程序 ………………………………………………… 041
4.3 上机实验 ………………………………………………………………………… 042
    4.3.1 实验 1：韩信点兵 ……………………………………………………… 042
    4.3.2 实验 2：水仙花数 ……………………………………………………… 043
4.4 拓展练习 ………………………………………………………………………… 044

第 5 章 Java 方法 …………………………………………………………………… 048

5.1 知识提炼 ………………………………………………………………………… 048
    5.1.1 方法的定义 ……………………………………………………………… 048
    5.1.2 方法的调用 ……………………………………………………………… 049
    5.1.3 方法的参数 ……………………………………………………………… 050
    5.1.4 方法的重载 ……………………………………………………………… 050
    5.1.5 方法的递归 ……………………………………………………………… 051
5.2 实例解析 ………………………………………………………………………… 052
5.3 上机实验 ………………………………………………………………………… 055
5.4 拓展练习 ………………………………………………………………………… 057

第 6 章 Java 数组 …………………………………………………………………… 059

6.1 知识提炼 ………………………………………………………………………… 059
    6.1.1 数组概述 ………………………………………………………………… 059
    6.1.2 一维数组 ………………………………………………………………… 059
    6.1.3 二维数组 ………………………………………………………………… 060
    6.1.4 数组的空间开辟 ………………………………………………………… 062
    6.1.5 数组工具类 Arrays 类 …………………………………………………… 063
6.2 实例解析 ………………………………………………………………………… 065
    6.2.1 实例 1：学生成绩等级判断 …………………………………………… 065
    6.2.2 实例 2：矩阵转置 ……………………………………………………… 066
6.3 上机实验 ………………………………………………………………………… 067

| | 6.4 | 拓展练习 | 069 |

第7章 类和对象 071

- 7.1 知识提炼 071
  - 7.1.1 类 071
  - 7.1.2 构造方法 071
  - 7.1.3 对象的创建与使用 072
  - 7.1.4 this 关键字 073
  - 7.1.5 static 关键字 075
- 7.2 实例解析 076
  - 7.2.1 实例1：复数类 076
  - 7.2.2 实例2：银行账户类 080
- 7.3 上机实验 085
- 7.4 拓展练习 089

第8章 继承与多态 092

- 8.1 知识提炼 092
  - 8.1.1 继承 092
  - 8.1.2 子类构造方法 092
  - 8.1.3 多态 094
- 8.2 实例解析 095
  - 8.2.1 实例1：银行信用卡类 095
  - 8.2.2 实例2：几何图形类 099
- 8.3 上机实验 103
- 8.4 拓展练习 108

第9章 接口与包 112

- 9.1 知识提炼 112
  - 9.1.1 接口 112
  - 9.1.2 接口的实现 112
  - 9.1.3 包 114
- 9.2 实例解析 115
  - 9.2.1 实例1：求平均值 115
  - 9.2.2 实例2：计算面积与体积 119
- 9.3 上机实验 122
- 9.4 拓展练习 126

第10章 文件读/写 130

- 10.1 知识提炼 130

　　　　10.1.1　File 类 …………………………………………………………… 130
　　　　10.1.2　字节流 ………………………………………………………… 130
　　　　10.1.3　字符流 ………………………………………………………… 131
　　　　10.1.4　随机访问文件 ………………………………………………… 131
　　　　10.1.5　对象序列化 …………………………………………………… 132
　　10.2　实例解析 …………………………………………………………………… 132
　　　　10.2.1　实例 1：完全数文件读/写 …………………………………… 132
　　　　10.2.2　实例 2：电话号码提取 ……………………………………… 134
　　　　10.2.3　实例 3：字数统计 …………………………………………… 136
　　　　10.2.4　实例 4：素数的随机读/写 …………………………………… 140
　　10.3　上机实验 …………………………………………………………………… 142
　　10.4　拓展练习 …………………………………………………………………… 146

第 11 章　泛型与集合 ………………………………………………………………… 149

　　11.1　知识提炼 …………………………………………………………………… 149
　　　　11.1.1　泛型 …………………………………………………………… 149
　　　　11.1.2　泛型类的子类及有界类型参数 ……………………………… 150
　　　　11.1.3　Collection ……………………………………………………… 151
　　　　11.1.4　Set 接口及其实现类 ………………………………………… 152
　　　　11.1.5　List 接口及其实现类 ………………………………………… 152
　　　　11.1.6　Map 接口及其实现类 ………………………………………… 153
　　11.2　实例解析 …………………………………………………………………… 154
　　　　11.2.1　实例 1：List 集合的基本使用 ………………………………… 154
　　　　11.2.2　实例 2：Map 集合的基本使用 ……………………………… 155
　　11.3　上机实验 …………………………………………………………………… 157
　　11.4　拓展练习 …………………………………………………………………… 160

第 12 章　GUI 编程 …………………………………………………………………… 162

　　12.1　知识提炼 …………………………………………………………………… 162
　　　　12.1.1　GUI 概述 ……………………………………………………… 162
　　　　12.1.2　容器组件 ……………………………………………………… 162
　　　　12.1.3　基本组件 ……………………………………………………… 163
　　　　12.1.4　布局管理器 …………………………………………………… 163
　　　　12.1.5　事件处理 ……………………………………………………… 163
　　　　12.1.6　菜单、其他组件 ……………………………………………… 164
　　12.2　实例解析 …………………………………………………………………… 165
　　　　12.2.1　实例 1：JList 列表 …………………………………………… 165
　　　　12.2.2　实例 2：求三角形面积 ……………………………………… 166
　　12.3　上机实验 …………………………………………………………………… 170

12.3.1　实验1：对键盘每个操作的监控 ················· 170
　　　12.3.2　实验2：设计留言板 ························· 171
　12.4　拓展练习 ····································· 177

第13章　JDBC编程 ································· 178

　13.1　知识提炼 ····································· 178
　　　13.1.1　数据库管理系统 ·························· 178
　　　13.1.2　JDBC的概念 ····························· 178
　　　13.1.3　JDBC API ······························· 178
　　　13.1.4　JDBC数据库连接的基本步骤 ················ 179
　13.2　实例解析 ····································· 180
　13.3　上机实验 ····································· 185
　13.4　拓展练习 ····································· 200

第14章　网络编程 ································· 202

　14.1　知识提炼 ····································· 202
　　　14.1.1　网络编程基本概念 ························ 202
　　　14.1.2　两类传输协议：TCP和UDP ················· 202
　　　14.1.3　URL的组成与创建 ························ 203
　　　14.1.4　InetAddress类 ··························· 204
　　　14.1.5　Socket通信原理 ·························· 204
　　　14.1.6　Applet对URL访问 ······················· 204
　14.2　实例解析 ····································· 204
　14.3　上机实验 ····································· 209
　14.4　拓展练习 ····································· 213

第 1 章

Java 入门

1.1 知识提炼

1.1.1 Java 运行平台

随着网络的飞速发展，Java 已经成为网络时代最重要的语言之一，主要优势体现在"一次编写，到处运行(write once, run anywhere)"，这需要提供相应的运行平台，目前 Java 运行平台主要有 3 个版本。

- J2SE(Java 2 Platform Standard Edition)包含构成 Java 语言核心的类，如数据库连接、接口定义、输入/输出和网络编程。主要用于开发一般个人计算机上的应用软件，后更名为 Java SE。
- J2ME(Java 2 Platform Micro Edition)包含 J2SE 中一部分类，主要用于消费类电子产品的软件开发，如呼机、智能卡、手机、PDA 和机顶盒，后更名为 Java SE。
- J2EE(Java 2 Platform Enterprise Edition)包含 J2SE 中的所有类，并且还包含用于开发企业级应用的类，如 EJB、Servlet、JSP、XML 和事务控制，也是现在 Java 应用的主要方向。主要用于开发企业级应用软件，后更名为 Java SE。

1.1.2 Java 程序的运行机制

在 Java 中源文件名称的后缀为.java，之后通过编译生成一个扩展名为.class 的与平台无关的字节码 class 文件，然后由 Java 虚拟机(JVM)解释执行。基本原理如图 1-1 所示。

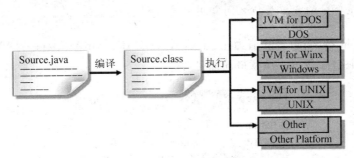

图 1-1 JVM 基本原理

在 Java 中源文件名称的后缀为.java，之后通过编译使 *.java 的文件生成一个 *.class 文件，在计算机上执行 *.class 文件，但是此时执行 *.class 的计算机并不是一个物理上可以看到的计算机,而是Java自己设计的一个计算机——JVM,所有的*.class文件都是在JVM上运行的,即*.class文件只需要认识JVM,由JVM再去适应各个操作系统。如果不同的操作系统安装上符合其类型的JVM,那么以后程序无论到哪个操作系统上都是可以正确执行的。Java也是通过JVM进行可移植性操作的。

1.1.3 Java SE 开发环境

一台计算机上安装了JVM即可运行Java程序,但是要开发Java程序,还需建立Java开发环境。不同领域的Java开发应用所需的版本不同,本书使用Java SE的开发环境。安装与配置开发环境的步骤如下。

步骤一:安装JDK(本书使用JDK 1.7版本)。

下载安装文件,按照安装向导提示一步步安装完成即可。

步骤二:设置环境变量。

在JDK安装完毕需设置path和classpath两个环境变量,这两个环境变量的设置非常关键,是程序编译和运行的重要保证。path指示java命令的路径,像javac、java、javaw等,这样在控制台下面编译、执行程序时就不需要再输入具体路径了。classpath是类库的默认搜索路径,即告诉JVM要使用或者执行的*.class文件所在的目录。这个是专门针对Java的,故系统里没有这条路径,即告诉JVM要使用或者执行的*.class文件所在的目录。由于JDK的安装路径多次使用,在此先新建环境变量JAVA_HOME。

(1) 新建环境变量JAVA_HOME,其值为C:\Java\jdk1.7.0_03,如图1-2所示。

(2) 设置变量Path,在变量值最前面增%JAVA_HOME%\bin;,如图1-3所示。

图1-2 设置JAVA_HOME路径

图1-3 设置Path路径

注意:环境变量的各变量值之间需用分号分隔。

(3) 新建变量classpath,其值为.;%JAVA_HOME%\lib\dt.jar;%JAVA_HOME%\lib\tools.jar,如图1-4所示。注意最前面是".;"。

图1-4 设置classpath路径

1.1.4 Java 开发工具

Java 的开发工具很多,目前比较流行的 Java 开发工具有 EditPlus、Jcreator、Eclipse、MyEclipse、Jbuilder、NetBeans 等,本书主要用的开发工具是 MyEclipse。MyEclipse 的功能非常强大,支持也十分广泛,尤其是对各种开源产品的支持。MyEclipse 可以支持 Java Servlet、AJAX、JSP、JSF、Struts、Spring、Hibernate 及 EJB3、JDBC 数据库链接工具等多项功能。图 1-5 为 MyEclipse 工作界面。

图 1-5　MyEclipse 工作界面

常用操作如下。

(1) 新建 Java Project:File→New→Java Project,如图 1-6 和图 1-7 所示。

(2) 新建各类对象:包(Package)、类(Class)等。

① 新建类的操作:选中相应的 Project 的 src 目录,右击 New→Class,如图 1-8 和图 1-9 所示。

② 新建包的操作:选中相应的 Project 的 src 目录,右击 New→Package,如图 1-10 所示,输入相应的包的名称即可,注意包的名称用小写字母,可以用点作为分隔符。本实验的代码均放在 gdlgxy.shiyan1 包中。

(3) 如果有些窗口不小心关闭了,可以使用 Show View 功能,如图 1-11 所示。

(4) 代码格式化,如图 1-12 所示,使用快捷键 Ctrl+Shift+F。

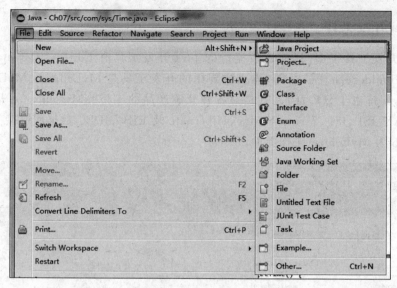

图 1-6 新建 Project(1)

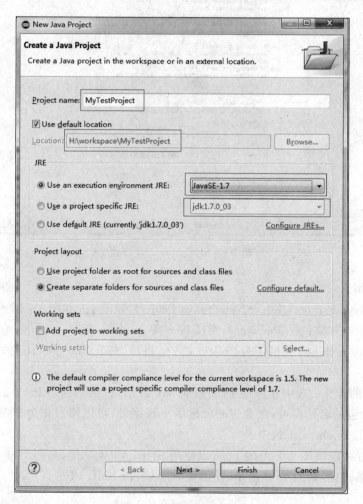

图 1-7 新建 Project(2)

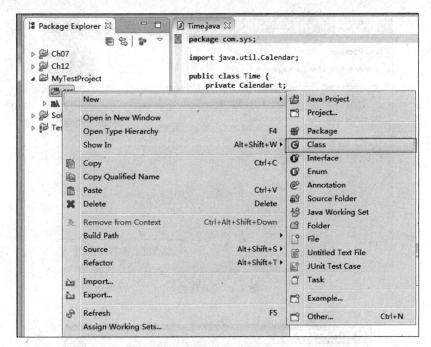

图 1-8 新建 Class(1)

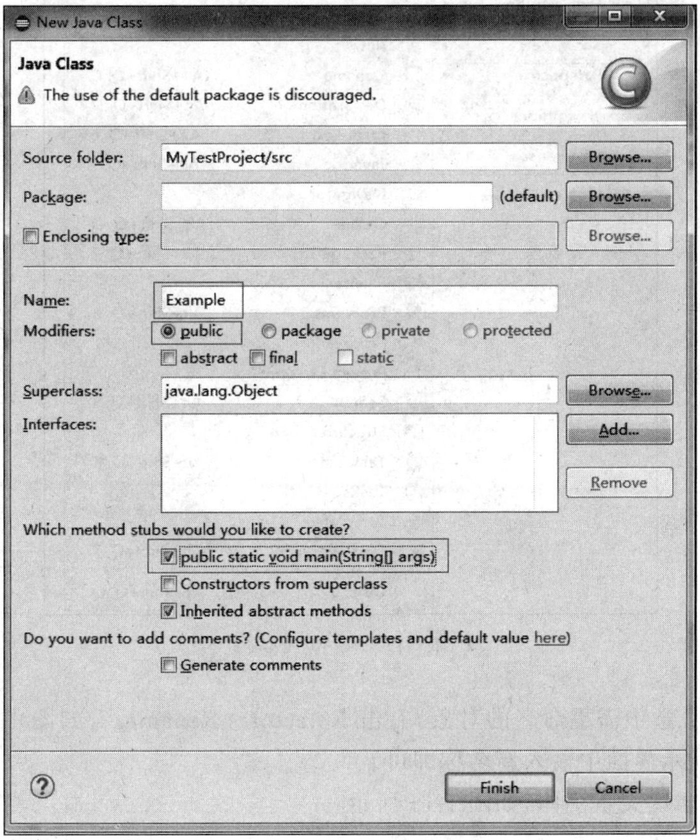

图 1-9 新建 Class(2)

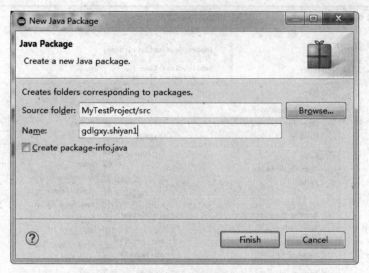

图 1-10　新建包

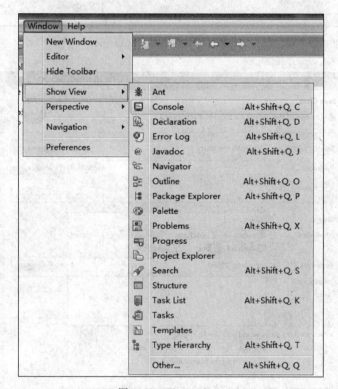

图 1-11　Show View

（5）重命名：选中需重命名的对象，右击 Refactor→Rename...，如图 1-13 和图 1-14 所示。在图 1-14 所示界面中输入新名称即可。

（6）修改编辑区文本字体，如图 1-15 所示。

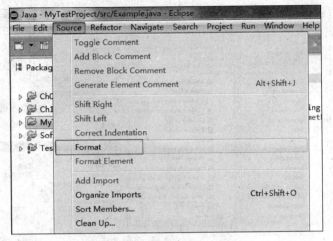

图 1-12　代码格式化

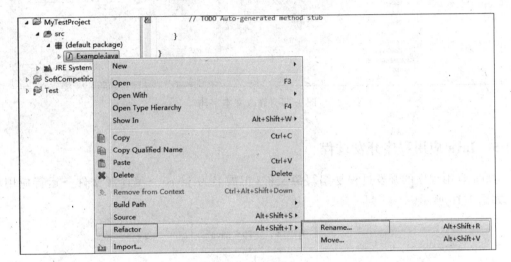

图 1-13　重命名(1)

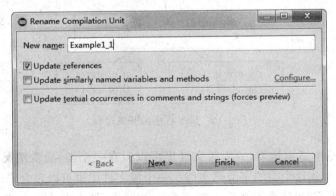

图 1-14　重命名(2)

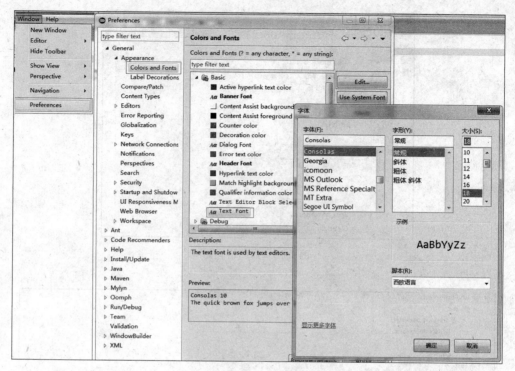

图 1-15　修改文本字体

1.1.5　Java 应用程序开发过程

Java 应用程序的开发过程是编写源文件(扩展名为.java)→编译源文件→运行应用程序,如图 1-16 所示。

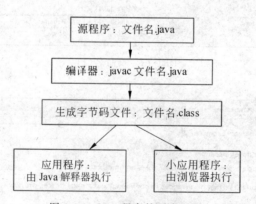

图 1-16　Java 程序的开发过程

Java 程序的基本结构是类,一个源文件里可以只有一个类,通常情况下源文件名即为类名,编译源文件得到这个类的字节码文件。一个源文件中也可以包含多个类,其中最多只能有一个 public 类,如果这个源文件中有一个 public 类,那么源文件名必须为 public 类的类名,编译源文件后会得到多个字节码文件,即每个类会生成一个字节码文件。

1.2 实例解析

【实例要求】

编写一个简单问答程序：系统提示输入姓名，用户输入姓名，系统输出欢迎信息，如图 1-17 所示。

请输入您的姓名：张三丰
张三丰欢迎您！
祝您学有所成！

图 1-17 简单问答程序

【实例源代码】

新建源文件 ShiyanDemo1.java，源代码如下：

```java
package gdlgxy.shiyan;
import java.util.Scanner;                          //导入 Scanner 类

public class ShiyanDemo1 {
    /* 功能描述：系统提示输入姓名，用户输入姓名，系统输出欢迎信息
    */
    public static void main(String[] args) {
        Scanner sc=new Scanner(System.in);         //构造 Scanner 类对象 sc
        System.out.print("请输入您的姓名:");        //提示输入姓名
        String name=sc.next();                     //接收,用户输入赋值给字符串变量 name
        System.out.println(name+"欢迎您!\n 祝您学有所成!"); //输出欢迎信息
    }
}
```

【实例解析】

(1) MyEclipse 的使用方法：新建 Java 项目、新建包、新建类。

(2) 读入数据。

在 java.util 包中，可使用 Scanner 类创建一个对象，实现数据的输入。因此，在前面需要用到 import 语句导入 Scanner 类。读入数据的方法有：

- nextInt()——读入整数。
- nextFloat()——读入浮点数。
- next()、nextLine()——读入字符串。

注意：next() 一定要读取到有效字符后才结束输入，对输入有效字符之前遇到的空格键、Tab 键或 Enter 键等结束符，next() 方法会自动将其去掉，只有在输入有效字符之后，next() 方法才将其后输入的空格键、Tab 键或 Enter 键等视为分隔符或结束符。next() 查找并返回再次扫描下一个完整标记。完整标记的前后是与分隔模式匹配的输入信息，所以 next() 方法不能得到带空格的字符串。而 nextLine() 方法的结束符只是 Enter 键，即 nextLine() 方法返回的是 Enter 键之前的所有字符，它是可以得到带空格的字符串的。

(3) 输出数据。

- println()方法输出信息后换行。

```
System.out.println();                    //输出换行符
System.out.println("XXX");               //输出信息并换行
```

- print()方法输出信息不换行。

```
System.out.print("XXX");
```

- printf()方法格式化输出。

```
System.out.printf("格式控制符",参数 1,参数 2,...,参数 n);    //格式化输出
```

其中,格式控制符字符串由普通字符和格式控制符组成,普通字符原样输出,格式控制符用以控制后面的参数以何种格式输出,后面的参数个数与格式控制符格式一致。格式控制符如下。

%d：输出 int 型数据。

%md：输出占 m 列 int 型数据。

%f：输出 float、double 浮点数。

%.nf：输出小数保留 n 位的浮点数。

%m.nf：输出占 m 列小数保留 n 位的浮点数。

%e：以指数形式输出 float、double 浮点数。

%c：输出 char 型数据。

%s：输出 String 型数据。

(4) 养成良好的编程习惯,如缩进、写注释等,掌握单行注释和多行注释方法。

1.3 上机实验

【实验目的】

- 了解常用的 Java 开发工具。
- 掌握安装并配置 Java 开发环境。
- 掌握 Java 的开发流程。
- 可以编写并运行一个简单的 Java 程序。

【实验要求】

使用 Java 语言完成一个简单的个性测试小程序。
依次提出以下问题,计算机给出判定回答。
你走到森林里,希望第一眼看见什么东西?
你走到森林里,希望第二眼看见什么东西?
继续往前走,看见一个屋子,你是绕一圈进去还是直接推门进去?答案请填写直接或者绕弯。

推开屋子门后,看见一个桌子,你希望是方的,还是圆的?答案请填写方的或者圆的。

桌子上面有张纸,纸上有个数字,10以内,你希望是几?

用户回答完以上问题,测试给出答案。

你的前生是(用户第一次输入的内容),你的另一半是(用户第二次输入的内容),你的性格是(用户第三次输入的内容和第四次输入的内容),你的幸运数字是(用户第五次输入的内容)。

实验运行效果如图1-18所示。

```
你走到森林里,希望第一眼看见什么东西?
小白兔
你走到森林里,希望第二眼看见什么东西?
小鹿
继续往前走。看见一个屋子,你是绕一圈进去还是直接推门进去?答案请填写直接或者绕弯
直接
推开屋子门后,看见一个桌子,你希望是方的,还是圆的?答案请填写方的或者圆的
圆的
桌子上面有张纸,纸上有个数字,10以内,你希望是几?
7
你的前生是:小白兔,你的另一半是:小鹿,你的性格是:直接圆的,你的幸运数字是:7
```

图1-18 实验运行效果

【实验指导】

本实验用到java.util包里的Scanner类的方法输入数据、java.lang包里的System.out.println方法输出数据,输出内容要使用字符串拼接方法。

【程序模板】

```java
package gdlgxy.shiyan;
import java.util.Scanner;
public class CharacterTest {
    public static void main(String[] args) {
        [代码1]    //构建Scanner对象sc
        System.out.println("你走到森林里,希望第一眼看见什么东西?");
        [代码2]    //接收第一个问题的答案
        System.out.println("你走到森林里,希望第二眼看见什么东西?");
        [代码3]    //接收第二个问题的答案
        System.out.println("继续往前走,看见一个屋子,你是绕一圈进去还是直接推门进去?答案请填写直接或者绕弯");
        [代码4]    //接收第三个问题的答案
        System.out.println("推开屋子门后,看见一个桌子,你希望是方的,还是圆的?答案请填写方的或者圆的");
        [代码5]    //接收第四个问题的答案
        System.out.println("桌子上面有张纸,纸上有个数字,10以内,你希望是几?");
        [代码6]    //接收第五个问题的答案
        sc.close();
    }
}
```

思考：测试给出的答案换用 System.out.printf 方法应该如何输出？

1.4 拓展练习

【基础知识练习】

一、选择题

1. 当初 Sun 公司发展 Java 的原因是(　　)。
   A. 发展航空仿真软件　　　　　　B. 发展人工智能软件
   C. 发展消费性电子产品　　　　　D. 发展统计分析软件
2. Java 是从(　　)语言改进并重新设计的。
   A. Ade　　　　B. C++　　　　C. Pascal　　　　D. C
3. Java 因为(　　)快速发展而走红。
   A. 个人计算机与网络　　　　　　B. 游戏软件
   C. 系统软件　　　　　　　　　　D. 人工智能
4. Java 源程序经过编译器编译后生成的字节码文件的扩展名为(　　)。
   A. java　　　　B. html　　　　C. c　　　　D. class
5. Java 语言是(　　)。
   A. 面向问题的解释型高级编程语言　　B. 面向机器的低级编程语言
   C. 面向对象的解释型高级编程语言　　D. 面向过程的编译型高级编程语言
6. 在 JDK 环境下编译 Java 源程序使用的命令是(　　)。
   A. java　　　　B. javac　　　　C. tomcat　　　　D. jvm
7. Java 平台的无关性是通过(　　)实现的。
   A. Java 的编辑环境　　　　　　　B. Java 虚拟机
   C. UNIX 操作系统　　　　　　　　D. Windows 操作系统
8. 在编写 Java Applet 程序时，需在程序的开头输入(　　)。
   A. import java.awt.*;　　　　　B. import java.applet.Applet;
   C. import java.io.*;　　　　　D. import java.awt.Graphics;
9. 运行 Java 程序需要的工具软件所在的目录是(　　)。
   A. JDK 的 bin 目录　　　　　　　B. JDK 的 demo 目录
   C. JDK 的 lib 目录　　　　　　　D. JDK 的 jre 目录
10. CLASSPATH 中的"."的含义是指(　　)。
    A. 省略号　　　B. 当前目录　　　C. 所有目录　　　D. 上级目录

二、填空题

1. Java 程序分为_____和_____两种。
2. Java 源程序需要通过编译器编译成为_____文件(也称_____文件)，Java 虚拟机中的 Java 解释器负责将_____解释成为特定的机器码进行运行。
3. 一个 Java Application 源程序文件名为 MyJavaTest.java，如果使用 Sun 公司的 Java 开发工具 JDK 编译该源程序文件并使用其虚拟机运算这个程序的字节码文件，应该顺序执行如下两个命令：_____、_____。

【上机练习】

使用 Java 语言完成以下简单的求和运算的程序：输入两个数，输出两数之和。效果如图 1-19 所示。

```
请输入第一个数：15
请输入第二个数：18.795
15.00+18.80=33.80
```

图 1-19　求和运算

提示：输入的两个数为实数，用 double 类型；输出结果保留两位小数，可用 printf( ) 方法；控制格式符用%.2f。

第 2 章

Java 基础语法

2.1 知识提炼

2.1.1 变量与常量

1. 变量

在程序运行过程中可以发生改变的量称为变量，其命名规则为：除第一个单词首字母要小写外，其他每个单词首字母要大写。变量的分类如图 2-1 所示。

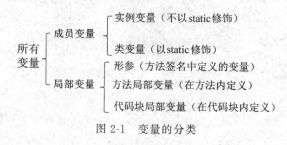

图 2-1 变量的分类

在方法或者语句块中声明的变量称为局部变量。例如，局部变量的定义：

```
Hvoid print(int i){              //方法的参数变量
    int j=2;                     //方法局部变量
    {
        int x=10;                //代码块局部变量
        int y=x+j;
    }
}
```

在方法体外，类体内声明的变量称为成员变量。例如，成员变量的定义：

```
private int a;                   //实例变量
public static float b;           //类变量
```

在定义上，成员变量定义时可用访问修饰符（public、private、protected、默认），局部变量定义时不能用访问修饰符。在引用上，引用实例变量方法为实例名.变量名，引用类变量方法为类名.变量名，对于局部变量的引用，在相应的位置直接使用变量名即可。

变量要先定义并初始化后才能参与运算。局部变量必须显示初始化，对于成员变量，如果没有显示初始化，Java 虚拟机会自动初始化为默认值。

（1）整数类型（byte、short、int、long）的基本类型变量的默认值为 0。

（2）单精度浮点型（float）的基本类型变量的默认值为 0.0f。

（3）双精度浮点型（double）的基本类型变量的默认值为 0.0d。

（4）字符型（char）的基本类型变量的默认值为\u0000。

（5）布尔型的基本类型变量的默认值为 false。

（6）引用类型的变量的默认值为 null。

（7）数组引用类型的变量的默认值为 null。如果数组变量的实例后没有显示为每个元素赋值，Java 就会把该数组的所有元素初始化为其相应类型的默认值。

成员变量的作用域是整个类内，而局部变量的作用域是在类中某个方法内或某段代码块内。

2. 常量

在程序运行过程中不能发生改变的量称为常量，在 Java 中用 final 修饰定义，一般常量用大写字母表示。声明常量时也可以不赋值，在需要时再赋值，但常量赋值只能有一次。

final 修饰的变量即常量，只能被赋值一次。例如，在一个类中定义常量：

```
public static final double PI=3.14;
```

2.1.2 基本数据类型

在 Java 中的数据类型分类如图 2-2 所示。

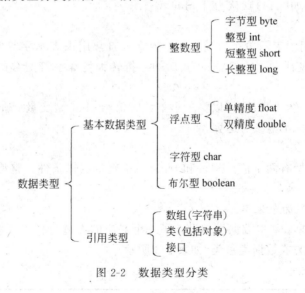

图 2-2　数据类型分类

1. 整数型

整数型用于表示没有小数部分的数值，它允许是负数。Java 提供了 4 种整型，具体内容如表 2-1 所示。

表 2-1 整数型

关键字	类 型	类型说明	长度(二进位)	取 值 范 围
byte	字节型	1字节	8	-128～127
short	短整型	2字节	16	-32768～32767
int	整型	4字节	32	-2147483648～2147483647
long	长整型	8字节	64	-9223372036854775808～9223372036854775807

一个整数(如1)默认数据为 int 型。在实际使用中要注意各整型的范围,Java 中数据超出了范围系统不会报错,而是输出一个不正确的结果。在 Java 中,当整数范围达到相应数据类型的最大值时,再加1就会变成最小值,故在定义一个整型变量的数据类型时,要先估算其值的取值范围,再指定其类型。

2. 浮点型

浮点型用于表示有小数部分的数值。在 Java 中有两种浮点型,具体内容如表 2-2 所示。

表 2-2 浮点型

关键字	类 型	类型说明	长度(二进位)	取 值 范 围
float	单精度浮点型	4字节	32	$\pm(1.4E-45)\sim\pm(3.4028235E+38)$(有效位数为6～7位)
double	双精度浮点型	8字节	64	$\pm(4.9E-324)\sim\pm(1.7976931348623157E+308)$(有效位数为15位)

一个浮点数值(如 3.14)默认为 double 型。

3. char 型

char(字符)型用于表示单个字符(占2字节),通常用来表示字符常量,用单引号引起来。Java 中还允许使用转义字符\来将其后的字符转为特殊字符型常量。常用的转义字符如\n 表示换行。

char 与 int 有天然的联系,可以将一个整数赋值给 char 型变量。char 型是可以进行运算的,会自动转换成 int 型参与运算。

4. boolean 型

boolean(布尔)型有两个值:false 和 true,用来判定逻辑条件。整型值和布尔值之间不能进行相互转换。

5. 基本数据类型的包装类

在设计类时为每个基本数据类型设计了一个对应的类进行代表,称为包装类(Wrapper Class),也称为外覆类或数据类型类,如表 2-3 所示。

表 2-3 包装类对应表

基本数据类型	包装类	基本数据类型	包装类
byte	Byte	int	Integer
boolean	Boolean	long	Long
short	Short	float	Float
char	Character	double	Double

6. 基本数据类型的转换

基本数据类型的转换分为自动转换和强制转换。图 2-3 给出了数值类型之间的自动转换。图中有 5 个实心箭头，表示无信息丢失的转换；有 3 个虚箭头，表示可能有精度损失的转换。

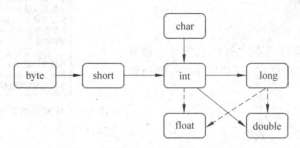

图 2-3　数据类型之间的自动转换

有多种类型的数据混合运算时，系统首先自动将所有数据转换成容量最大的那种数据类型，然后再进行计算。byte、short、char 之间不会相互转换，他们三者在计算时首先转换为 int 类型。

自动类型转换的逆过程将容量大的数据类型转换为容量小的数据类型，即强制转换。使用时要加上强制转换符()，但可能造成精度降低或溢出，需要格外注意。强制类型转换的语法格式是在括号中给出想要转换的目标类型，后面紧跟待转换的变量名。

把字符串转换为数值的方法是 Xxx.parseXxx(String)的形式，其中 Xxx 对应不同的数值类型。把数值转换为字符串的方法是 String.valueOf()。字符串使用双引号引起来，要与字符型区分开来。

例如，编写声明不同数据类型变量的程序文件源代码如下：

```java
public class Example2_1 {
    public static void main(String args[]) {
        byte b=19;
        short s=0x19ff;                    //十六进制数
        int i=1000000;
        long l=0xffffL;
        char c='F';
        float f=0.19F;
        double d=0.19E-3;                  //科学计数法 0.19 * 10-3
        boolean boo=false;
        String str="我爱 Java";String str="这是字符串类数据类型";
        System.out.println("字节型变量 b ="+b);
        System.out.println("短整型变量 s ="+s);
        System.out.println(" 整型变量 i ="+i);
        System.out.println("长整型变量 l ="+l);
        System.out.println("字符型变量 c ="+c);
        System.out.println("浮点型变量 f ="+f);
        System.out.println("双精度变量 d ="+d);
        System.out.println("布尔型变量 boo ="+boo);
        System.out.println("字符串类对象 str ="+str);
```

```
        }
    }
```

编译并运行该程序,结果如图 2-4 所示。

2.1.3 运算符

在 Java 中的运算符基本上可以分为算术运算符、赋值运算符、关系运算符、逻辑运算符、位运算符和其他运算符等。

```
字节型变量 b = 19
短整型变量 s = 6655
整型变量 i = 1000000
长整型变量 l = 65535
字符型变量 c = F
浮点型变量 f = 0.19
双精度变量 d = 1.9E-4
布尔型变量 boo = false
字符串类对象 str = 我爱Java
```

图 2-4 基本数据类型演示程序运行结果

1. 算术运算符

Java 的算术运算符如表 2-4 所示。

表 2-4 算术运算符

运算符	运算	范例	结果
+	正号	+3	3
-	负号	b=4;-b	-4
+	加	5+5	10
-	减	6-4	2
*	乘	3*4	12
/	除	5/5	1
%	取模	7%5	2
++	自增(前):先运算后取值 自增(后):先取值后运算	a=2;b=++a; a=2;b=a++;	a=3;b=3 a=3;b=2
--	自减(前):先运算后取值 自减(后):先取值后运算	a=2;b=--a; a=2;b=a--;	a=1;b=1 a=1;b=2
+	字符串相加	"He"+"llo"	"Hello"

2. 赋值运算符

符号=为赋值运算符。当=两侧数据类型不一致时,可以使用自动类型转换原则或使用强制类型转换原则进行处理,支持连续赋值。例如,a=b=c=10;,即将三个变量赋值为 10。

3. 关系运算符

关系运算符如表 2-5 所示,用于比较两个数值之间的大小,其运算结果为一个逻辑类型(boolean 型)的值。

表 2-5 关系运算符

运算符	运算	范例	结果
==	相等于	4==3	false
!=	不等于	4!=3	true
<	小于	4<3	false
>	大于	4>3	true

续表

运算符	运算	范例	结果
<=	小于等于	4<=3	false
>=	大于等于	4>=3	true

要注意区别==与=这两个运算符。

4．逻辑运算符

逻辑运算符要求操作数的数据类型为 boolean 型，其运算结果也是 boolean 型。逻辑运算符有：&—逻辑与；|—逻辑或；!—逻辑非；&&—短路与；||—短路或；^—逻辑异或。

逻辑运算符的真值表如表 2-6 所示，其中 A 和 B 是逻辑运算的两个逻辑变量。

表 2-6 逻辑运算符的真值表

A	B	A&&B	A||B	!A	A^B	A&B	A|B
false	false	false	false	true	false	false	false
true	false	false	true	false	true	false	true
false	true	false	true	true	true	false	true
true	true	true	true	false	false	true	true

在程序设计时使用 && 和 || 运算符，可提高运算效率，不建议使用 & 和 | 运算符。

5．位运算符

位运算是以二进制位为单位进行的运算，其操作数和运算结果都是整型值，如表 2-7 所示。

表 2-7 位运算符

运算符	运算	范例
<<	左移	3<<2=12，即 3*2*2=12
>>	右移	3>>1=1，即 3/2=1
>>>	无符号右移	3>>>1=1，即 3/2=1
&	与运算	6&3=2
|	或运算	6|3=7
^	异或运算	6^3=5
~	反码	~6=-7

位运算的位与&、位或|、位非~、位异或^与逻辑运算的相应操作的真值表完全相同（false 表示 0，true 表示 1），其差别只是位运算操作的操作数和运算结果都是二进制整数，而逻辑运算相应操作的操作数和运算结果都是逻辑值 boolean 型。

6．条件运算符（三目运算符）

<表达式 1>? <表达式 2>：<表达式 3>

说明：先计算<表达式 1>的值，当<表达式 1>的值为 true 时，则将<表达式 2>的值作为整个表达式的值；当<表达式 1>的值为 false 时，则将<表达式 3>的值作为整个表

达式的值。

7. new 运算符与点运算符

new 运算符用于构建一个类的对象(实例),后面接该类的构造方法。

点运算符.的功能有两个:一是引用类中成员;二是指示包的层次等级。两者后面将详述。

例如,编写使用关系运算符和逻辑运算符的程序文件,源代码如下:

```java
public class Example2_2 {
    public static void main(String[] args) {
        int a = 25, b = 20, e = 3, f = 0;
        boolean d = a < b;
        System.out.println("a=25,b=20,e=3,f=0");
        System.out.println("因为关系表达式 a<b 为假,所以其逻辑值为: " + d);
        if (e != 0 && a / e > 5)
            System.out.println("因为 e 非 0 且 a/e 为 8 大于 5,所以输出 a/e= " + a / e);
        if (f != 0 && a / f > 5)
            System.out.println("a/f = " + a / f);
        else
            System.out.println("因为 f 值为 0,所以输出 f = " + f);
    }
}
```

编译并运行该程序,结果如图 2-5 所示。

```
a=25,b=20,e=3,f=0
因为关系表达式 a<b 为假,所以其逻辑值为 : false
因为e非0且a/e为8大于5,所以输出 a/e= 8
因为f值为0,所以输出 f=0
```

图 2-5 常用运算符演示程序运行结果

2.1.4 语句与复合语句

Java 中一条完整的语句是以分号结束的,而多条语句用花括号括起来称为复合语句。

例如,编写包含复合语句程序,源代码如下:

```java
class Example2_3{
    public static void main(String args[]) {
        int k, i=3, j=4;
        k=i+j;
        System.out.println("在复合块外的输出 k="+k);
        {   //复合语句开始
            float f;
            f=j+4.5F;
            i++;
            System.out.println("在复合块内的输出 f="+f);
            System.out.println("在复合块内的输出 k="+k);
        }   //复合语句结束
```

```
            System.out.println("在复合块外的输出 i="+i);
        }
    }
```

编译并运行该程序，结果如图 2-6 所示。

```
在复合块外的输出  k-7
在复合块内的输出  f=8.5
在复合块内的输出  k=7
在复合块外的输出  i=4
```

图 2-6 复合语句演示程序运行结果

2.2 实例解析

【实例要求】

编程求一个 4 位整数的逆序数加 1，例如，输入 1234，输出 4322，如图 2-7 所示。

```
请输入一个4位整数:
1234
1234的逆序数加1为: 4322
```

图 2-7 求逆序数

【实例源代码】

新建源文件 ShiyanDemo2.java，源代码如下：

```
package gdlgxy.shiyan;
import java.util.Scanner;
public class ShiyanDemo2 {
    public static void main(String[] args) {
        Scanner sc =new Scanner(System.in);
        System.out.println("请输入一个 4 位整数:");
        int num =sc.nextInt();          //接收第一个问题的答案
        int ge=num%10;
        int shi=num/10%10;
        int bai=num/100%10;
        int qian=num/1000%10;
        int revnum=ge * 1000+shi * 100+bai * 10+qian+1;
        System.out.println(num+"的逆序数加 1 为:"+revnum);
    }
}
```

【实例解析】

整个过程是先拆后装，通过算术运算符分别获得各个位上的数字，从左向右得到第 n 位（n 从 0 开始，即 n=0 表示个位，n=1 表示十位……）上的数字公式为：$num/10^n\%10$。例如，取个位：$ge=num/10^0\%10=num\%10$；取十位：$shi=num/10^1\%10=num/10\%10$……然后再重新得到其逆序数，个位变成千位，十位变成百位……以此类推。

此例也可用到 StringBuffer 类的 reverse() 方法得到反转的字符串，String 类与 int 型的转换方法如图 2-8 所示。

```
Scanner sc=new Scanner(System.in);
System.out.print("请输入一个整数：");
int x=sc.nextInt();
StringBuffer s=new StringBuffer(String.valueOf(x));
String s1=s.reverse().toString();
int y=Integer.parseInt(s1)+1;
System.out.println(x+"的逆序数+1为："+y);
System.out.println("班级：16软件4班");
System.out.println("学号：01");
System.out.println("姓名：张三");
```

图 2-8　使用 reverse() 方法转换

2.3　上机实验

【实验目的】

- 了解 Java 标识符与关键字。
- 掌握各种变量的声明方式。
- 理解运算符的优先级。
- 掌握 Java 基本数据类型的使用。
- 掌握运算符的使用方法。

【实验要求】

编写一个简易复数计算器程序：分别输入两个复数的实部和虚部，输出它们的和、差、积、商。

实验运行效果如图 2-9 所示。

```
======简易复数运算器======
请输入第一个数的实部和虚部（用空格分隔）：
-2.5 -7.9
第一个数为：-2.50-7.90i
请输入第二个数的实部和虚部（用空格分隔）：
-1 5.68
第二个数为：-1.00+5.68i
======运算结果如下：
(-2.50-7.90i)+(-1.00+5.68i)=-3.50-2.22i
(-2.50-7.90i)-(-1.00+5.68i)=-1.50-13.58i
(-2.50-7.90i)*(-1.00+5.68i)=47.37-6.30i
(-2.50-7.90i)/(-1.00+5.68i)=-1.27+0.66i
```

图 2-9　实验运行效果

【实验指导】

(1) 复数四则运算法则。

假设 $z1=a+bi, z2=c+di$

和：$z1+z2=(a+c)+(b+d)i$

差：$z1-z2=(a-c)+(b-d)i$

积：$z1*z2=(ac-bd)+(bc+ad)i$

商：$z1/z2=(ac+bd)/(c*c+d*d)+(bc-ad)/(c*c+d*d)i$

（2）复数表达式方法。

这里要多次获得复数的表达式，故在类中定义了一个静态方法(类方法)toString()。

```
public static String toString(double a,double b){//复数表达式
    if(b==0) return "不是复数";
    String x=b>0? String.format("%.2f+%.2fi",a,b):
                  String.format("%.2f-%.2fi",a,-b);
    return x;
}
```

用三元运算符区分当虚部为正负两种情况的表示方法。用String.format()方法使得各部保留2位小数。

【程序模板】

```
package gdlgxy.shiyan;
import java.util.Scanner;
public class ComplexTest {
    public static void main(String[] args) {
        Scanner sc=new Scanner(System.in);
        System.out.println("======简易复数运算器======");
        System.out.println("请输入第一个数的实部和虚部(用空格分隔):");
        double a=sc.nextDouble();
        double b=sc.nextDouble();
        String x=toString(a,b);         //第一个复数 x
        System.out.println("第一个数为:"+x);
        System.out.println("请输入第二个数的实部和虚部(用空格分隔):");
        double c=[代码 1];
        double d=[代码 2];
        String y=[代码 3];               //第二个复数 y
        System.out.println("第二个数为:"+[代码 4]);
        double suma=a+c;                //和实部
        double sumb=b+d;                //和虚部
        String sum=toString(suma,sumb);//和
        double mnsa=[代码 5];            //差实部
        double mnsb=[代码 6];            //差虚部
        String mns=toString(mnsa,mnsb);//差
        [代码 7];                        //积实部
        [代码 8];                        //积虚部
        [代码 9];                        //积
        double t=c*c+d*d;
```

```
            double diva=(a*c+b*d)/t;        //商实部
            double divb=(b*c-a*d)/t;        //商虚部
            String div=toString(diva,divb);//商
            System.out.println("======运算结果如下:");
            System.out.printf("(%s)+(%s)=%s\n",x,y,sum);
            [代码 10];
            [代码 11];
            [代码 12];
        }
        public static String toString(double a,double b){   //复数表达式
            String x=b>0? String.format("%.2f+%.2fi",a,b):
                         String.format("%.2f-%.2fi",a,-b);
            return x;
        }
    }
```

思考：如果得到复数的虚部为 0，应该如何处理，即如何修改 toString()方法处理 b==0 这种情况？

2.4 拓展练习

【基础知识练习】

一、选择题

1. 下面( )是合法的 Java 标识符。
   A. #_pound   B. _underscore   C. 5Interstate   D. class
2. 下面( )是 Java 语言中的关键字。
   A. sizeof   B. abstract   C. NULL   D. Native
3. 表达式(11+3*8)/4%3 的值是( )。
   A. 31   B. 0   C. 1   D. 2
4. 下列运算符中，优先级别最高的是( )。
   A. %   B. &&   C. >   D. =
5. 设 x=1,y=2,z=3,则表达式 y+=z--/++x 的值是( )。
   A. 3   B. 3.5   C. 4   D. 5
6. Java 中定义常量的保留字是( )。
   A. const   B. final   C. finally   D. native
7. Java Application 程序中需要 main()方法，main()方法的返回值类型是( )。
   A. int   B. void   C. 空   D. double
8. 为一个 boolean 类型变量赋值时，可以使用( )方式。
   A. boolean = 0;
   B. boolean a = ((9-4)>=10);
   C. boolean a="true";

D. boolean a＝＝false；

9. 有 a、b 两个变量的定义如 int a＝2；int b＝3；。下列表达式的值为 true 的是(　　)。
   A．a＋＋＝＝b－－　　　　　　　　B．＋＋a＝＝b－－
   C．＋＋a＝＝－－b　　　　　　　　D．a＋＋＝＝b

10. 下列(　　)属于引用数据类型。
    A．String　　　　B．char　　　　C．double　　　　D．int

二、填空题

1. 设 x＝2，则表达式(x＋＋)＊3 的值是＿＿＿＿＿＿＿。
2. 若 x＝6，y＝12，则 x＞＝y 和 x＜＝y 的逻辑值分别是＿＿＿＿＿＿和＿＿＿＿＿＿。
3. 假设 x＝13，y＝6，则表达式 x％y!＝0 的值是＿＿＿＿＿＿，其数据类型是＿＿＿＿＿＿。
4. 已知语句 int a＝10,b＝0,c；if(a＜50){b＝9；}c＝b＋a；System.out.println(c)；的值是＿＿＿＿＿＿。
5. 已知语句 int a＝7；System.out.println(a％3)；的值是＿＿＿＿＿＿。
6. 已知语句 int a＝11；System.out.println(a/2)；的值是＿＿＿＿＿＿。
7. Java 中每个数据均赋予一种数据类型，1 默认数据类型为＿＿＿＿＿＿，1.0 默认数据类型为＿＿＿＿＿＿。

【上机练习】

输入一个秒数，要求转换为××小时××分××秒的格式输出。运行结果如图 2-10 所示。

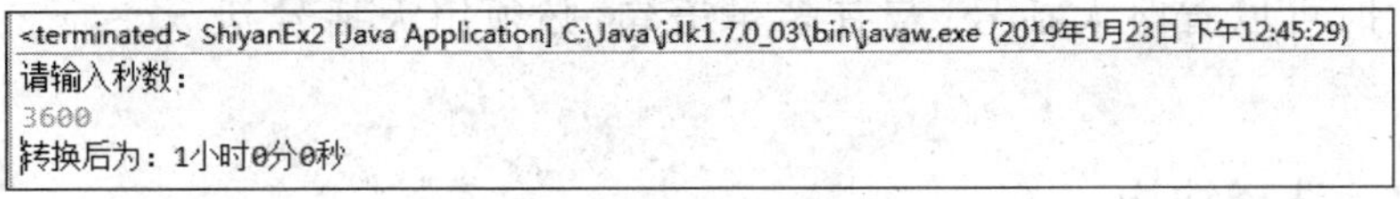

```
<terminated> ShiyanEx2 [Java Application] C:\Java\jdk1.7.0_03\bin\javaw.exe (2019年1月23日 下午12:45:29)
请输入秒数：
3600
转换后为：1小时0分0秒
```

图 2-10　运行结果

第 3 章

选 择 结 构

3.1 知识提炼

3.1.1 if 选择结构

只有 if 的结构是单分支结构,程序流程图如图 3-1 所示。
语法格式如下:

```
if(条件表达式){
    代码块
}
```

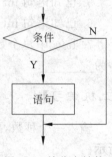

图 3-1 单分支结构

说明:条件表达式必须是一个布尔表达式,必须用圆括号括起来,如果条件中的值为 true 就执行代码块,否则跳过。代码块可以是一条语句,此时可以省略大括号;若有多条语句,必须用大括号括起来。

3.1.2 if-else 选择结构

最常用的分支语句是 if-else 双分支结构,程序流程图如图 3-2 所示。

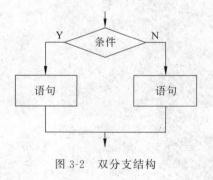

图 3-2 双分支结构

语法格式如下:

```
if(条件表达式){
    代码块 1
}
```

```
else{
    代码块 2
}
```

说明：如果条件表达式中的值为 true，执行代码块 1；否则执行代码块 2。

3.1.3 多重 if 选择结构

多重 if 结构构成了多分支结构，程序流程图如图 3-3 所示。

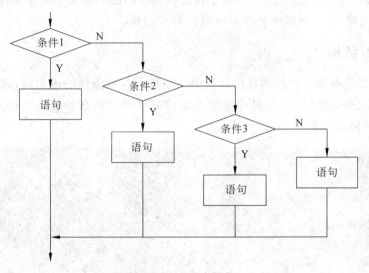

图 3-3 多分支结构

语法格式如下：

```
if(条件表达式 1){
    代码块 1
}else if(条件表达式 2){
    代码块 2
}else{
    代码块 3
}
```

说明：解决需要判断的条件是连续的区间时有很大优势，else if 块可以有多个，取决于程序的需要，如果条件表达式 1 为 true，执行代码块 1，否则执行 else if 块；如果条件表达式 2 为 true，执行代码块 2，否则执行代码块 3，以此类推；当条件满足某个 else if 块时，则余下的将不再执行而跳出 if 块。

此外，可在 if 或者 else 里嵌套 if 结构，语法格式如下：

```
if(条件表达式 1){                    if(条件表达式 1){
    if(条件表达式 2){                    代码块 1
        代码块 1                     }else{
```

```
        }else{                              if(条件表达式 2){
            代码块 2                             代码块 2
        }                                   }else{
    }else{                                      代码块 3
        代码块 3                             }
    }                                   }
```

说明：左边的程序是在 if 选择结构里嵌入 if 选择结构,条件表达式 1 为 false 时执行代码块 3,否则执行内部 if 选择结构,也就是说要执行代码块 1,则必须满足条件表达式 1 及条件表达式 2。同理,右边的程序是在 else 里嵌套 if 结构。

3.1.4　switch 结构

要在许多的选择条件中找到并执行其中一个符合判断条件的语句时,除了可以使用多重 if-else 判断之外,还可以使用另一种更方便的方式即多重选择——switch 语句,也称为开关语句,语法格式如下:

```
switch(表达式){
    case 常量 1:
        代码块 1;
        break;
    case 常量 2:
        代码块 2;
        break;
    ...
    default:
        码块 3;
}
```

说明：

（1）switch 后面的表达式类型可以是 byte、short、char、int、枚举、String(不能为浮点型和 long 型)。

（2）case 后面的值不能相同,否则会出现互相矛盾的情况。

（3）default 块在其他 case 块都不满足的情况下执行,default 块是可有可无的,如果它不存在,并且所有的常量值都和表达式的值不同,那么该语句就不进行任何处理。

（4）break 语句用来在执行完一个 case 分支后使程序跳出 switch 语句块；如果没有 break 语句,程序会顺序执行到 switch 结尾。

（5）与嵌套 if 选择结构相比,switch 选择结构方便于解决等值判断问题。

3.1.5　if 与 switch 的比较

if 和 switch 语句很像,具体应用到哪些场景呢？

虽然这两个语句都可以使用,但如果判断的具体数值不多,而且符合 byte、short、int、char 这四种类型,建议使用 switch 语句,因为效率较高。对区间的判断、对结果为 boolean 型判断,建议使用 if 语句,因为 if 的使用范围更广。

例如，用 switch 代码实现在不同温度时显示不同的解释说明，代码如下：

```
int c=38;
switch (c<10? 1:c<25? 2:c<35? 3:4) {
    case 1:
        System.out.println(" "+c+"℃ 有点冷。要多穿衣服。");
        break;
    case 2:
        System.out.println(" "+c+"℃ 正合适。出去玩吧。");
        break;
    case 3:
        System.out.println(" "+c+"℃ 有点热。");
        break;
    default:
        System.out.println(" "+c+"℃ 太热了！开空调。");
}
```

也可用 if 语句实现，代码如下：

```
int c=38;
if(c<10)
    System.out.println(" "+c+"℃ 有点冷。要多穿衣服。");
else if(c<25)
    System.out.println(" "+c+"℃ 正合适。出去玩吧。");
else if(c<35)
    System.out.println(" "+c+"℃ 有点热。");
else
    System.out.println(" "+c+"℃ 太热了！开空调。");
```

编译运行程序，其结果如图 3-4 所示。

38℃ 太热了！开空调。

图 3-4 分支结构演示程序运行结果

3.2 实例解析

【实例要求】

编程：系统提示输入学生成绩，用户输入成绩，系统输出成绩等级（90分以上为 A，80~89 分为 B，60~79 分为 C，0~59 分为 D），如图 3-5 所示，要求分别用 if-else、switch 选择结构语句实现。

【实例源代码】

（1）用 if-else 语句实现。

新建源文件 ShiyanDemoIf3.java，源代码如下：

| 请输入学生成绩：200 | 请输入学生成绩：-10 | 请输入学生成绩：95 |
| 输入成绩不正确！ | 输入成绩不正确！ | 成绩等级为A |

| 请输入学生成绩：80 | 请输入学生成绩：76 | 请输入学生成绩：38 |
| 成绩等级为B | 成绩等级为C | 成绩等级为D |

图 3-5　简单问答程序

```java
package gdlgxy.shiyan;
import java.util.Scanner;
public class ShiyanDemoIf3 {
    /* 功能描述:用 if-else 语句实现成绩等级判断 */
    public static void main(String[] args) {
        Scanner sc=new Scanner(System.in);
        System.out.print("请输入学生成绩:");
        int score=sc.nextInt();    //输入成绩
        String degree;
        if(score<0 || score>100) degree="输入成绩不正确!";
        else if(score>=90)
            degree="成绩等级为 A";
        else if(score>=80)
            degree="成绩等级为 B";
        else if(score>=60)
            degree="成绩等级为 C";
        else
            degree="成绩等级为 D";
        System.out.println(degree);
        sc.close();
    }
}
```

（2）用 switch 语句实现。

新建源文件 ShiyanDemoSwitch3.java，源代码如下：

```java
package gdlgxy.shiyan;
import java.util.Scanner;
public class ShiyanDemoSwitch3 {
    /* 功能描述:用 Switch 语句实现成绩等级判断 */
    public static void main(String[] args) {
        Scanner sc=new Scanner(System.in);
        System.out.print("请输入学生成绩:");
        int score=sc.nextInt();
        String degree =null;
        int d=score/10;
        if(score<0 || score>100) degree="输入成绩不正确!";
        else {
            switch(d){
                case 10:
```

```
                case 9:
                    degree="成绩等级为 A"; break;
                case 8:
                    degree="成绩等级为 B"; break;
                case 7:
                case 6:
                    degree="成绩等级为 C"; break;
                case 5:
                case 4:
                case 3:
                case 2:
                case 1:
                case 0:
                    degree="成绩等级为 D"; break;
            }
        }
        System.out.println(degree);
        sc.close();
    }
}
```

【实例解析】

(1) 分数应为 0~100 分,须排除不合法数据:if(score<0 || score>100),条件表达式中使用"||"运算符表示两个条件中满足其一则表示数据不合法。

(2) ShiyanDemoIf3.java 中,使用 if-else 语句判断成绩的等级。

(3) ShiyanDemoSwitch3.java 中,switch 后的表达式如果直接使用 score,则需要用 101 条分支,即 case 0、case 1、…、case 100,显然使用这种方式不是很好。本例中将成绩整除 10,即 int d=score/10,d 的范围为 0~10,此时,只需用到 11 条分支,即 case 0、case 1、…、case 10,从而使代码简洁,效率高。

(4) 在 switch 语句中,也可连续写下一系列 case,默认后面的语句块,以指定多种情况下运行相同的语句块,这时最后一个 case 代码块适用于前面的所有 case。本例代码中,d 为 10、9 时,成绩等级为 A;d 为 5、4、3、2、1、0 时,成绩等级为 D。此外,本例 switch 语句中默认使用了 default,因为之前使用了 if-else 语句使得成绩范围为 0~100 分,因此 d 的取值范围为 0~10,也可使用 default 语句,代码如下:

```
switch(d){
    case 10:
    case 9:
        degree="成绩等级为 A"; break;
    case 8:
        degree="成绩等级为 B"; break;
    case 7:
```

```
        case 6:
            degree="成绩等级为 C"; break;
        default:
            degree="成绩等级为 D";
    }
```

3.3 上机实验

【实验目的】

- 理解 Java 程序语法结构。
- 掌握 if-else 语句的使用。
- 掌握 switch 语句的使用。
- 熟悉 break 语句的作用。

【实验要求】

分别用 if-else、switch 语句模拟系统菜单选择程序。
实验运行效果如图 3-6 所示。

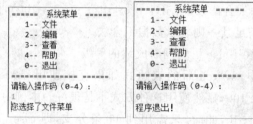

图 3-6　实验运行效果

【实验指导】

本实验用到 java.util 包里的 Scanner 类的方法输入数据、java.lang 包里的 System.out. println()方法输出数据，输出内容要使用字符串拼接方法。

【程序模板】

```java
package gdlgxy.shiyan;
import java.util.Scanner;
public class MenuChooseTest {
    public static void main(String[] args) {
        menuShow();                          //显示系统菜单
        System.out.println("请输入操作码(0-4):");
        Scanner sc=new Scanner(System.in);
```

```
        [代码1]    //输入操作码
        sc.close();
        if(op==0){                       //操作码为0,程序退出
            System.out.println("程序退出!");
            System.exit(0);
        }
        [代码2] //if...else语句块,输入不同操作码执行不同操作
    }
    public static void menuShow(){        //定义系统菜单,并输出
        System.out.println("======  系统菜单   ======");
        System.out.println("    1--文件         ");
        System.out.println("    2--编辑         ");
        System.out.println("    3--查看         ");
        System.out.println("    4--帮助         ");
        System.out.println("    0--退出         ");
        System.out.println("=====================");
    }
}
```

思考：程序中也可不定义 menuShow() 方法，将方法体中的代码直接放到 main() 方法中，应该如何修改？

3.4 拓展练习

【基础知识练习】

一、选择题

1. 复合语句用(　　)括起来的一段代码。
   A. 小括号()   B. 大括号{}   C. 中括号[]   D. 单引号''
2. 下列不属于条件语句关键字的是(　　)。
   A. if   B. else   C. switch   D. while
3. 多分支语句 switch(表达式){}中,表达式不可以返回(　　)的值。
   A. 整型   B. 实型   C. 接口型   D. 字符型
4. 下面不属于 Java 条件分支语句结构的是(　　)。
   A. if 结构   B. if-else 结构   C. if-else if 结构 D. if-else else 结构
5. 三元条件运算符 ex1？ex2：ex3,相当于下面(　　)语句。
   A. if(ex1) ex2;else ex3;   B. if(ex2) ex1;else ex3;
   C. if(ex1) ex3;else ex2;   D. if(ex3) ex2;else ex1;

二、填空题

1. 以下程序段的输出结果是_____。

```
int x=5,y=6,z=4;
if(x+y>z && x+z>y && z+y>x)
```

```
        System.out.println("三角形");
else
        System.out.println("不是三角形");
```

2. 以下程序段的输出结果是_____。

```
int i=1;
switch(i){
    case 0:
        System.out.print ("zero");
    case 1:
        System.out.print ("one");
    case 2:
        System.out.print ("two");
    default:
        System.out.print ("default");
}
```

3. 以下程序的输出结果为_____。

```
public class Test1{
    public static void main(String[] args) {
        String s1="hello";
        String s2=new String("hello");
        if(s1.equals(s2)){
            System.out.println("相等");}
        else {
            System.out.println("不相等");}
    }
}
```

4. 以下程序的输出结果是_____。

```
public class Test4 {
    public static void main(String[] args) {
        int x=-1;
        int y=4;
        int k;
        if((k=x++)<=0&&!(y--<=0)){
            k++;
        }else{
            k--;
        }
        System.out.println(x+","+y+","+k);
    }
}
```

5. 以下程序的输出结果是_____。

```java
public class Test10 {
    public static void main(String[] args) {
        int i=10;
        int j=11;
        int a=12;
        int b=13;
        if(i++>=j--&&++a>=b++){
            System.out.println("i="+i+" j="+j);
        }else{
            System.out.println("a="+a+" b="+b);
        }
    }
}
```

【上机练习】

用 Math 类的 random() 方法产生一个字符,输出该字符。若该字符是一个英文字母,则输出"生成的是英文字母";若该字符是一个数字,则输出"生成的是数字";否则输出"生成的是其他字符"。运行效果如图 3-7 所示。

```
生成的字符:`           生成的字符: F           生成的字符: d           生成的字符: 8
生成的是其他字符        生成的是英文字母        生成的是英文字母        生成的是数字
```

图 3-7　运行效果

提示:Math 类的 random() 方法产生的随机数在 0.0 和 1.0 之间,当该随机数乘以 128 后,其值在 0.0 和 128.0 之间,将它转换为 char 型后,用 if 来判断。

第 4 章 循环结构

4.1 知识提炼

4.1.1 while 语句

while 循环语句属于当型循环,程序流程图如图 4-1 所示。

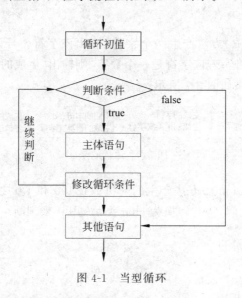

图 4-1 当型循环

语法格式如下:

```
while (条件表达式) {
    循环体
}
```

说明:

(1) 在执行时,先判断条件表达式,如果条件表达式的结果为 true,则执行循环体。当执行完循环体后,while 语句再判断条件表达是否成立,如果为 true,则继续下一轮循环。如此循环往复,直到条件表达式的结果为 false 时,循环体会被跳过,而执行 while 循环之后的语句。

(2) 循环体包含一条或多条语句,当只有一条语句时可以省略大括号。最后通常是更改循环条件的语句。此外,循环体也可为空语句。

(3) 如果一开始条件表达式的值为 false,则循环一次也不执行。

4.1.2 do-while 语句

do-while 循环语句属于直到型循环,程序流程图如图 4-2 所示。

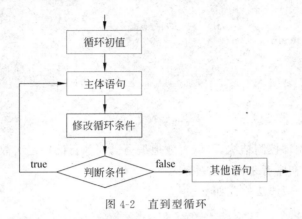

图 4-2 直到型循环

语法格式如下:

```
do {
    循环体
}while(条件表达式);
```

说明:do-while 循环也是用于未知循环执行次数的时候,而 while 循环与 do-while 循环的最大不同就是进入 while 循环前,while 语句会先测试判断条件的真假,再决定是否执行循环主体,而 do-while 循环则是"先做再说",每次都是先执行一次循环主体,然后再测试判断条件的真假,所以无论循环成立的条件是什么,使用 do-while 循环时,至少都会执行一次循环主体。

例如,求 1+2+…+100 之和,并将求和表达式与所求的和显示出来。运行结果如下:

```
1+2+...+100=5050
```

使用 do-while 语句实现的代码如下:

```java
class Example4_1 {
    public static void main(String args[]) {
        int n=1, sum=0;
        do {
            sum+=n++;
        }while (n<=100);
        System.out.println("1+2+...+100 ="+sum);
    }
}
```

使用 while 语句实现的代码如下：

```
class Example4_2 {
    public static void main(String args[]) {
        int n=1, sum=0;
        while (n<=100) {
            sum+=n++;
        }
        System.out.println("1+2+...+100 ="+sum);
    }
}
```

4.1.3 for 语句

for 循环是一个循环控制结构，可以有效地编写需要执行的特定次数的循环，for 语句简洁、使用率高，但对于初学者来说有一些难度。

for 循环语句的一般语法格式如下：

```
for(变量初始化;条件表达式;循环变量更新) {
    循环体
}
```

在 for 语句中，变量初始化部分只在开始时执行一次，然后判断条件表达式为 true，则执行循环体，再执行循环变量更新，接着再判断条件表达式是否成立，以决定一下次循环；若条件表达式为 false，则结束整个循环语句。因此，for 与 while 语句一样，如果首次执行条件表达式不成立，则循环体一次都不执行。其执行过程如图 4-3 所示。

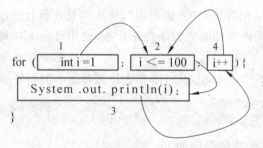

图 4-3 for 语句执行过程

上例中求 1～100 的累加和用 for 语句实现的代码如下：

```
class Example4_3 {
    public static void main(String args[]) {
        int sum=0;
        for(int n=1;n<=100;n++) {
            sum+=n;
        }
        System.out.println("1+2+...+100 ="+sum);
```

```
    }
}
```

4.1.4　break 与 continue 语句

break 语句有 3 种作用：①在 switch 语句中,它被用来终止一个语句序列。②在循环语句中,它被用来退出当前循环。③它能作为一种"先进"goto 语句来使用,即 break 后接语句标号。

break 语句可以强迫程序中断循环,当程序执行到 break 语句时会离开循环,继续执行循环外的下一个语句,如果 break 语句出现在嵌套循环中的内层循环,则 break 语句只会跳出当前层的循环。

continue 语句可以强迫程序跳到循环的起始处,当程序运行到 continue 语句时,即会停止运行剩余的循环主体语句,而回到循环的开始处进入执行下一次循环。

例如,比较以下两段代码的输出结果有什么不同？

代码一：

```java
public class TestBreak{
    public static void main(String args[]){
    for(int i =1 i<10; i++){
        if(i%3==0)
            break;
        System.out.print (i);
    }
    System.out.println("Game Over!");
    }
}
```

输出 1 和 2。

代码二：

```java
public class ContinueTest {
    public static void main(String args[]){
        for (int i =1; i <10; i++) {
            if (i%3==0) continue;
            System.out.print(i);
        }
    }
}
```

输出 10 以内不能被 3 整除的数：1、2、4、5、7、8。

4.1.5　多重循环

如果要输出二维表格。例如乘法表,则需要二重循环结构,即在一个循环结构语句的循环体内,再嵌入另一个循环语句。二重以上的循环就是多重循环。

前面介绍的几种循环语句可以相互嵌套,比如 for 语句里可以嵌套 while、do-while 或者 for 语句。

4.2 实例解析

4.2.1 实例 1:累加程序

【实例要求】

编写计算 $1 \sim n$ 的累加程序,运行结果如图 4-4 所示。

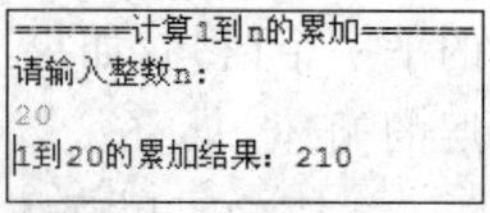

【实例源代码】

新建源文件 ShiyanDemoSum4.java,源代码如下:

```
======计算1到n的累加======
请输入整数n:
20
1到20的累加结果: 210
```

图 4-4　求累加和程序运行结果

```java
package gdlgxy.shiyan;
/* 功能描述:求 1 到 n 的累加和 */
    import java.util.Scanner;
public class ShiyanDemoSum4 {
    public static void main(String[] args) {
        int s=0;
        Scanner sc=new Scanner(System.in);
        System.out.println("=====计算 1 到 n 的累加=====");
        System.out.print("请输入整数 n:");
        int n=sc.nextInt();
        int i=1;
        while(i<=n){
            s+=i;
            i++;
        }
        System.out.printf("1到%d的累加结果:%d\n", n,s);
        sc.close();
    }
}
```

【实例解析】

本例中使用 while 语句实现求 $1 \sim n$ 的累加和,和 s 初始值应置为 0,加数 i 初始值为 1,每执行一次循环则 i 自增 1,循环体执行了 n 次,直到 $i = n + 1$ 结束循环。也可用 for 语句实现,代码如下:

```java
for(int i=1;i<=n;i++){
    s+=i;
}
```

4.2.2 实例 2：乘法口诀程序

【实例要求】

输出乘法口诀表，如图 4-5 所示。

1 * 1= 1	1 * 2= 2	1 * 3= 3	1 * 4= 4	1 * 5= 5	1 * 6= 6	1 * 7= 7	1 * 8= 8	1 * 9= 9
2 * 1= 2	2 * 2= 4	2 * 3= 6	2 * 4= 8	2 * 5=10	2 * 6=12	2 * 7=14	2 * 8=16	2 * 9=18
3 * 1= 3	3 * 2= 6	3 * 3= 9	3 * 4=12	3 * 5=15	3 * 6=18	3 * 7=21	3 * 8=24	3 * 9=27
4 * 1= 4	4 * 2= 8	4 * 3=12	4 * 4=16	4 * 5=20	4 * 6=24	4 * 7=28	4 * 8=32	4 * 9=36
5 * 1= 5	5 * 2=10	5 * 3=15	5 * 4=20	5 * 5=25	5 * 6=30	5 * 7=35	5 * 8=40	5 * 9=45
6 * 1= 6	6 * 2=12	6 * 3=18	6 * 4=24	6 * 5=30	6 * 6=36	6 * 7=42	6 * 8=48	6 * 9=54
7 * 1= 7	7 * 2=14	7 * 3=21	7 * 4=28	7 * 5=35	7 * 6=42	7 * 7=49	7 * 8=56	7 * 9=63
8 * 1= 8	8 * 2=16	8 * 3=24	8 * 4=32	8 * 5=40	8 * 6=48	8 * 7=56	8 * 8=64	8 * 9=72
9 * 1= 9	9 * 2=18	9 * 3=27	9 * 4=36	9 * 5=45	9 * 6=54	9 * 7=63	9 * 8=72	9 * 9=81

图 4-5　乘法口诀表（1）

【实例源代码】

新建源文件 ShiyanDemoKoujue4.java，源代码如下：

```java
package gdlgxy.shiyan;
import java.util.Scanner;
public class ShiyanDemoKoujue4{
    /* 功能描述:输出乘法口诀表 */
    public static void main(String[] args) {
      int k;
        for(int i=1;i<10;i++){
        for(int j=1;j<10;j++){
          k=i*j;
          System.out.printf("%d * %d=%2d", i,j,k);
        }
        System.out.println();            //输出换行
        }
    }
}
```

【实例解析】

乘法口诀表可看作一个二维表格，因此需要用二重循环实现。这里使用 for 语句里嵌套 for 语句，外循环中循环变量 i 控制行，内循环中循环变量 j 控制列。如果要输出三角形乘法口诀表，则代码如下：

```java
for(int i=1;i<10;i++){
    for(int j=1;j<=i;j++){
      k=i*j;
      System.out.printf("%d * %d=%2d", j,i,k);
    }
```

```
        System.out.println();                    //输出换行
    }
```

输出结果如图 4-6 所示。

```
1 * 1= 1
1 * 2= 2  2 * 2= 4
1 * 3= 3  2 * 3= 6  3 * 3= 9
1 * 4= 4  2 * 4= 8  3 * 4=12  4 * 4=16
1 * 5= 5  2 * 5=10  3 * 5=15  4 * 5=20  5 * 5=25
1 * 6= 6  2 * 6=12  3 * 6=18  4 * 6=24  5 * 6=30  6 * 6=36
1 * 7= 7  2 * 7=14  3 * 7=21  4 * 7=28  5 * 7=35  6 * 7=42  7 * 7=49
1 * 8= 8  2 * 8=16  3 * 8=24  4 * 8=32  5 * 8=40  6 * 8=48  7 * 8=56  8 * 8=64
1 * 9= 9  2 * 9=18  3 * 9=27  4 * 9=36  5 * 9=45  6 * 9=54  7 * 9=63  8 * 9=72  9 * 9=81
```

图 4-6 乘法口诀表(2)

4.3 上机实验

【实验目的】

- 理解 Java 程序语法结构。
- 掌握 while、do-while 语句的使用。
- 掌握 for 语句的使用。
- 熟悉 break、continue 语句的作用。

4.3.1 实验1：韩信点兵

【实验要求】

韩信在一次点兵的时候,为了不让敌人知道自己部队的军事实力,采用了下述点兵方法:先令士兵 1～3 报数;又令士兵 1～5 报数;后又令士兵 1～7 报数。韩信根据每次最后一个士兵报的数字很快算出自己部队士兵的总数。请编程计算韩信共有多少士兵。

```
士兵1~3报数,最后一个士兵报数为: 1
士兵1~5报数,最后一个士兵报数为: 2
士兵1~7报数,最后一个士兵报数为: 4
总人数为: 67
```

实验运行效果如图 4-7 所示。

图 4-7 实验运行效果

【实验指导】

士兵 1～3 报数,结果最后一个士兵报 x,说明士兵总人数是 3 的倍数余 x;士兵 1～5 报数,结果最后一个士兵报 y,说明士兵总人数是 5 的倍数余 y;士兵 1～7 报数,结果最后一个士兵报 z,说明士兵总人数是 7 的倍数余 z。因此,求整除 3 余 $x(x<3)$、整除 5 余 $y(y<5)$、整除 7 余 $z(z<7)$ 的最小自然数。

输入 3 个非负整数 a、b、c,表示每种队形排列的人数($a<3$、$b<5$、$c<7$)。例如,输入: 1 2 4。

输出总人数的最小值(或报告无解,即输出 No answer)。例如,输出: 67。

【程序模板】

```
package gdlgxy.shiyan;
import java.util.Scanner;
public class HanXinDianBing {
    public static void main(String[] args) {
        Scanner sc=new Scanner(System.in);
        int num;
        System.out.print("士兵1~3报数,最后一个士兵报数为:");
        int x=sc.nextInt();
        System.out.print("士兵1~5报数,最后一个士兵报数为:");
        [代码1]
        System.out.print("士兵1~7报数,最后一个士兵报数为:");
        [代码2]
        for([代码3]){
            if([代码4])    break;
        }
        System.out.print("总人数为:"+num);
        sc.close();
    }
}
```

思考：程序中用到了循环结构,有没有算法效率更高的算法？（**提示**：不用循环结构。）

我国古时候流传着一种算法,但这种算法的名称很多,宋朝周密叫它"鬼谷算",又名"隔墙算"；杨辉叫它"剪管术"；而比较通行的名称是"韩信点兵"。最初记述这类算法的书是《孙子算经》,后来在宋朝经过数学家秦九韶的推广,又发现了一种算法,叫作"大衍求一术"。这在数学史上是极有名的问题,外国人一般把它称为"中国剩余定理"。至于它的算法,在《孙子算经》上就已经有了说明,而且后来还流传着这么一道口诀：三人同行七十稀,五树梅花廿一枝,七子团圆正半月,除百零五便得知。

这就是韩信点兵的计算方法,它的意思是：凡是用 3 个一数剩下的余数 x,将它用 70 去乘（因为 70 是 5 与 7 的倍数,而又是以 3 去除余 1 的数）；5 个一数剩下的余数 y,将它用 21 去乘（因为 21 是 3 与 7 的倍数,而又是以 5 去除余 1 的数）；7 个一数剩下的余数 z,将它用 15 去乘（因为 15 是 3 与 5 的倍数,而又是以 7 去除余 1 的数）,将这些数加起来,若超过 105 就减去 105,如果剩下的数目还是比 105 大,就再减去 105,直到比 105 小为止。这样,所得的数就是士兵的总数。根据这个道理,列出算式：

$$x \times 70 + y \times 21 + z \times 15 - 105$$

4.3.2 实验 2：水仙花数

【实验要求】

输出所有的水仙花数。
实验运行效果如图 4-8 所示。

153 370 371 407

图 4-8 实验运行效果

【实验指导】

所谓水仙花数,是指一个3位数,其各个位上数字立方和等于其本身。例如:
$$153 = 1×1×1 + 3×3×3 + 5×5×5$$
水仙花数是一个3位整型数,数的范围是[100,1000)。

【程序模板】

```
package gdlgxy.shiyan;
import java.util.Scanner;
public class Daffodil {
    public static void main(String[] args) {
        int i,j,k,s;
        for(s=100; [代码1] ;s++){
            i=s%10;                    //取个位
            j=[代码2];                 //取十位
            k=[代码3];                 //取百位
            if([代码4]){
                System.out.print(s+"  ");
            }
        }
    }
}
```

4.4 拓展练习

【基础知识练习】

一、选择题

1. 一个循环一般应包括( )内容。
   A. 初始化部分   B. 循环体部分
   C. 迭代部分和终止部分   D. 以上都是

2. 关于while和do-while循环,下列说法正确的是( )。
   A. 两种循环除了格式不同外,功能完全相同
   B. 与do-while语句不同的是,while语句的循环至少执行一次
   C. do-while语句首先计算终止条件,当条件满足时,才去执行循环体中的语句
   D. 以上都不对

3. 以下由do-wile语句构成的循环执行的次数是( )。
   A. 一次也不执行   B. 执行一次
   C. 无限次   D. 有语法错,不能执行

4. 以下选项中循环结构合法的是（　　）。

    A.
    ```
    while (int i<7){
        i++;
        System.out.println("i is"+i);
    }
    ```

    B.
    ```
    int j=3;
    while(j) {
        System.out.println("j is"+j);
    }
    ```

    C.
    ```
    int j=0;
    for(int k=0; j+k !=10; j++,k++) {
        System.out.println("j is"+j+"k is"+k);
    }
    ```

    D.
    ```
    int j=0;
    do{
        System.out.println("j is"+j++);
        if (j ==3) {continue loop; }
    }while (j<10);
    ```

5. 下列语句序列执行后，i 的值是（　　）。

```
int i=32;
do {
    i/=2;
} while( i<3 );
```

  A. 16   B. 8   C. 4   D. 2

二、填空题

1. 循环结构中，用于循环控制语句的有_____和_____。
2. 下列程序输出结果为_____。

```
public class Test {
    public static void main(String[] args) {
        int a=0;
        outer: for(int i=0;i<2;i++)   {
```

```
            for(int j=0;j<2;j++) {
                if(j>i){
                    continue outer;}
                a++;
            }
        }
        System.out.println(a);
    }
}
```

3. 下列程序输出结果是_____。

```
public class Test3 {
    public static void main(String[] args) {
        for(int x=0;x<=50;x+=10){
            //x 等于 30 时跳出循环
            if( x ==30 ) {
                break;
            }
            System.out.print( x +",");
        }
    }
}
```

4. 下列程序输出结果是_____。

```
public class Test4 {
    public static void main(String[] args) {
        for(int x=0;x<=50;x+=10 ) {
            if( x ==30 ) {
                continue;
            }
            System.out.print( x+" ");
        }
    }
}
```

【上机练习】

1. 输入起始年份,输出 100 年内的闰年,要求每行输出 10 个年份,并统计个数。运行结果如图 4-9 所示。

```
请输入起始年份: 2000
从2000开始, 100年以内的闰年有:
2000  2004  2008  2012  2016  2020  2024  2028  2032  2036
2040  2044  2048  2052  2056  2060  2064  2068  2072  2076
2080  2084  2088  2092  2096
从2000开始, 100年以内的闰年共有25个
```

图 4-9 统计闰年运行结果

提示：

（1）判断闰年的条件表达式为 i％4＝＝0 && i％100!＝0 || i％400＝＝0;。

（2）控制每输出 10 个年份需要换行,通过计数器控制：if(counter％10＝＝0) System.out.println();。

2. 有 1、2、3、4 四个数字,能组成多少个互不相同且一个数字中无重复数字的三位数?并把他们都输出。

第 5 章

Java 方法

5.1 知识提炼

5.1.1 方法的定义

方法定义包括方法声明和方法体,一般语法格式如下:

```
[<访问修饰符>][<修饰符>]<返回值类型><方法名>([参数列表]) [throws<异常类>] {
                                                                //方法声明
    方法体;
}
```

说明:

(1) 访问修饰符有 public、private、protected 和默认四种情况,访问权限如表 5-1 所示。

表 5-1 访问修饰符权限说明

访问修饰符名称	说 明	备 注
public(公共)	可以被所有类访问	
protected(受保护)	可以被同一包中的所有类访问,可以被所有子类访问	子类不在同一包中也可以访问
private(私有)	只能够被当前类的方法访问	
默认(无访问修饰符)	可以被同一包中的所有类访问	如果子类不在同一个包中,也不能访问

(2) 修饰符是可选项,主要的修饰符作用如表 5-2 所示。

表 5-2 修饰符说明

修饰符名称	说 明	备 注
static	静态方法(又称为类方法,其他的称为实例方法)	提供不依赖于类实例的服务,并不需要创建类的实例就可以访问静态方法
final	防止任何子类重载该方法	注意不要使用 const;可以同 static 一起使用,避免对类的每个实例维护一个副本
abstract	抽象方法,类中已声明而没有实现的方法	不能将 static()方法、final()方法或者类的构造器方法声明为 abstract

(3) 返回值类型:声明方法必须声明返回类型(构造方法除外),方法的返回值类型有 int、double、String 等。没有返回值的方法,必须给出返回类型 void,表示空类型。如果返回

值声明不是 void,则方法体中一定要有 return 语句。

（4）参数列表：参数列表是可选的。方法可以没有参数,也可以有多个。如果有多个,则用英文逗号分隔。每个参数都要声明数据类型。方法声明中的参数是没有确定值的,属于形式参数,简称形参。

5.1.2 方法的调用

方法只有被调用才会被执行。方法定义一次,允许多次调用。

1. 一个类内部的方法调用

前面提到在方法声明处使用 static 修饰的为静态方法(类方法),没有 static 修饰的是非静态方法(实例方法)。

类方法调用的语法格式如下：

```
方法名(实参列表);
//实参列表中参数个数、数据类型和次序必须和所调用方法的形式参数列表匹配
```

对于实例方法的调用分为如下两种情况。
（1）在静态方法(类方法)中调用实例方法的语法。
① 先实例化类(使用 new 创建一个对象)。
② 对象名.方法名(实参)。
（2）如果是实例方法中调用实例方法,则也直接使用方法名调用。

2. 不同类之间的方法调用

调用其他类的类方法的语法格式如下：

```
类名.方法名(实参列表);
```

与前面不同的是调用类方法时,前面要加上类名限定。

调用其他类的实例方法如下。
（1）先实例化类(使用 new 创建一个对象)。
（2）对象名.方法名(实参)。

例如,在下面代码中定义了一个类方法 getArea()求三角形的面积,定义了一个实例方法 getInfo()输出三角形的面积。在 main()方法中分别调用了这两个方法。

```
public class Example5_1 {
    public static double getArea(double a,double b,double c) {    //类方法
        double s,area;
        s = (a+b+c)/2;
        area =Math.sqrt(s * (s-a) * (s-b) * (s-c));
        return area;
    }
    public void getInfo(){                    //实例方法
        double s=getArea(3,4,5);              //调用类方法
        System.out.println("三条边分别为 3、4、5 的三角形的面积是:"+s);
    }
```

```
    public static void main(String[] args) {
        double area=getArea(3,6,7);          //调用类方法
        System.out.println("三条边分别为 3、6、7 的三角形的面积是:"+area);
        Example5_1 ex=new Example5_1();
        ex.getInfo();                        //调用实例方法
    }
}
```

运行结果如下：

三条边分别为 3、6、7 的三角形的面积是:8.94427190999916
三条边分别为 3、4、5 的三角形的面积是:6.0

5.1.3 方法的参数

参数是方法调用时进行信息交换的渠道之一。方法的参数可分为：形式参数和实际参数。形式参数简称形参，出现在方法定义中，在整个方法体内都可以使用，离开该方法则不能使用。实际参数简称实参，出现在主调方法中，进入被调方法后，实参变量也不能使用。

当对带参数的方法进行调用时，实际参数将会传递给形式参数，也就是实参与形参结合的过程。实参和形参的功能是数据传送。实参和形参在数量上、类型上、顺序上应严格一致，否则会发生"类型不匹配"的错误。

调用方法与被调用方法之间往往需要进行数据传送，数据传送的方式有值传送和引用传送两种。

值传送方式是将调用方法的实参的值计算出来赋予被调用方法对应形参的一种数据传送方式，值传送方式的特点是"数据的单向传送"。使用值传送方式时，形参一般是基本类型的变量；实参则是常量、变量，也可以是表达式。

使用引用传送方式时，方法的参数类型一般为复合类型（引用类型），复合类型变量中存储的是对象的引用。所以在参数传送中是传送引用，方法接收参数的引用，这样任何对形参的改变都会影响到对应的实参。因此，引用传送方式的特点是：引用的单向传送，数据的双向传送。

Java 中进行方法调用传递参数时，遵循传递的原则是：基本类型传递的是该数据本身，引用类型传递的是对象的引用，不是对象本身。

5.1.4 方法的重载

方法的重载是指在一个类中定义多个同名的方法，但要求每个方法具有不同的参数类型或参数个数。调用重载方法时，Java 编译器能通过检查调用的方法的参数类型和个数选择一个恰当的方法。方法重载通常用于创建完成一组任务相似但参数的类型或参数的个数不同的方法。方法重载是让类以统一的方式处理不同类型数据的一种手段。

注意：方法的重载与返回值类型无关，只看参数列表，且参数列表必须不同（参数个数或参数类型）。调用时，根据方法参数列表的不同来区别。

5.1.5 方法的递归

在 Java 中,递归是指方法在运行过程序中直接或间接调用自身而产生的重入现象。递归是一种思想,一个合法的递归定义包含两个部分:基础情况和递归部分。分析一个递归问题就是列出递归定义表达式的过程。

例如,求自然数的阶乘。

递归思想:

$$n! = n \times (n-1)!$$

首先分析表达式:

$$f(n) = \begin{cases} 1, & n=1 \\ f(n-1) \times n, & n>1 \end{cases}$$

求斐波那契数列第 n 项的递归算法如下:

```
long factorial(int n){
    if (n ==1) return 1;
    return factorial (n-1) * n;
}
```

从代码中可以看出,递归必须满足以下两个条件。

(1) 在每一次调用自身时,必须是(在某种意义上)更接近于解。

(2) 必须有一个终止处理或计算的准则。

执行过程如图 5-1 所示。

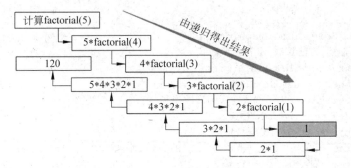

图 5-1 求 5 的阶乘递归执行过程

例如,求两个数的最大公约数的递归算法如下:

```
int gcd(int a,int b){
    return a%b==0? b:gcd(b,a%b);
}
```

在这个方法中出现了调用自身的现象,当 a%b==0 时,方法返回 b,终止计算。

5.2 实例解析

【实例要求】

(1) 在一个类中使用方法的重载编写求圆、长方形和三角形面积的方法,在 main() 方法中分别调用输出各图形的面积。运行结果如图 5-2 所示。

(2) 输出斐波那契数列的前 n 项,如图 5-3 所示。

图 5-2 求图形面积程序运行结果

图 5-3 斐波那契数列

【实例源代码】

(1) 求图形面积。

新建源文件 ShiyanDemoGraphicArea5.java,源代码如下:

```java
package gdlgxy.shiyan;
import java.util.Scanner;
public class ShiyanDemoGraphicArea5 {
    /* 功能描述:方法重载.求各图形面积 */
    public static void main(String[] args) {
        Scanner sc=new Scanner(System.in);
        System.out.print("请输入圆形的半径:");
        double r=sc.nextDouble();
        double area=getArea(r);
        System.out.printf("半径为%.2f的圆形的面积为 %.2f\n",r,area);
        System.out.println();
        System.out.print("请输入矩形的长和宽:");
        double len=sc.nextDouble();
        double wid=sc.nextDouble();
        area=getArea(len,wid);
        System.out.printf("长为%.2f、宽为%.2f的矩形的面积为 %.2f\n",len,wid,area);
        System.out.println();
        System.out.print("请输入三角形的三条边长:");
        double a=sc.nextDouble();
```

```
            double b=sc.nextDouble();
            double c=sc.nextDouble();
            area=getArea(a,b,c);
            System.out.printf("边长分别为%.2f、%.2f、%.2f的三角形的面积为 %.2f\n",a,b,c,area);
            sc.close();
    }
    //求圆形的面积
    public static double getArea(double r){
        double area=Math.PI * r * r;
        return area;
    }
    //求矩形的面积
    public static double getArea(double len,double wid){
        double area=len * wid;
        return area;
    }
    //求三角形的面积
    public static double getArea(double a,double b,double c){
        double s=(a+b+c)/2;
        double area=Math.sqrt(s * (s-a) * (s-b) * (s-c));
        return area;
    }
}
```

【实例解析】

本实例中使用了方法的重载,定义了三个同名的方法 getArea()。第一个 getArea()方法的功能是求圆的面积,该方法只有一个参数,即圆的半径;第二个 getArea()方法的功能是求矩形的面积,该方法有两个参数,即矩形的长和宽;第三个 getArea()方法的功能是求三角形的面积,该方法有三个参数,即三角形的三条边长。在 main()方法中三次调用 getArea()方法,根据实参的个数来决定调用的是哪个方法。

(2)输出斐波那契数列的前 n 项。

新建源文件 ShiyanDemoFibonacci5.java,源代码如下:

```
package gdlgxy.shiyan;
import java.util.Scanner;
public class ShiyanDemoFibonacci5 {
    /* 功能描述:斐波那契数列变量实现,求前n项,这种方法最简单,但是最占用内存
       特别是当n很大的时候效率低  */
    public static void main(String[] args) {
        Scanner sc=new Scanner(System.in);
        System.out.print("请输入 n:");
        int n=sc.nextInt();
        for (int i =1; i <=n; i++) {
            System.out.println("第" +i +"项" +getFib(i));
```

```
        }
        sc.close();
    }
    public static int getFib(int n) {                    //核心函数
        if (n < 0) {
            return -1;                                    //小于 0 是没有意义的
        } else if (n ==0) {
            return 0;                                     //等于 0 其实也没什么意义
        } else if (n ==1 || n ==2) {
            return 1;                                     //第一项和第二项都是 1
        } else {
            return getFib(n -1) +getFib(n -2);   //返回当前项,核心递归
        }
    }
}
```

【实例解析】

斐波那契数列的排列是：0,1,1,2,3,5,8,13,21,34,55,89,144……以此类推,即它后一个数等于前面两个数的和。

递归思想：一个数等于前两个数的和。

数列的递归表达式：

$$f(n)=\begin{cases}n, & n\leqslant 1\\ f(n-1)+f(n-2), & n>1\end{cases}$$

斐波那契数列执行过程如图 5-4 所示。

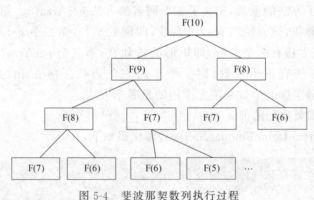

图 5-4 斐波那契数列执行过程

递归写法简单优美,省去考虑很多边界条件的时间。当然,递归算法会保存很多的临时数据,类似于堆栈的过程,如果栈深太深,就会造成内存用尽、程序崩溃的现象。Java 为每个线程分配了栈大小,如果栈大小溢出,就会报错,这时还是选择递归好一点。

观察图 5-4 所示的执行过程也会发现,本程序并没有保存每次的运算结果,第三行的 F(7) 就执行了两次,下层的 F(1)、F(2) 的次数更是以指数级增长。这也是本程序的一个弊端。

5.3 上机实验

【实验目的】

- 掌握方法的定义和调用。
- 掌握方法的重载的实现。
- 理解方法的递归思想。

【实验要求】

分别输入一元二次方程的二次项、一次项和常数项系数，输出一元二次方程的根。要求定义方法求方程的根。

实验运行效果如图 5-5 所示。

```
==============================
    1-- 一元二次方程求解
    2-- 一元一次方程求解
    0-- 退出
==============================
请输入操作代码：5
操作码输入错误，请重新输入！
==============================
    1-- 一元二次方程求解
    2-- 一元一次方程求解
    0-- 退出
==============================
请输入操作代码：1
请输入一元二次方程的二次项系数：4
请输入一元二次方程的一次项系数：5
请输入一元二次方程的常数项系数：6
一元二次方程无实根
==============================
    1-- 一元二次方程求解
    2-- 一元一次方程求解
    0-- 退出
==============================
请输入操作代码：2
请输入一元一次方程的一次项系数：4.5
请输入一元一次方程的常数项系数：2.8
一元一次方程根为：-0.62
==============================
    1-- 一元二次方程求解
    2-- 一元一次方程求解
    0-- 退出
==============================
请输入操作代码：0
程序结束！
```

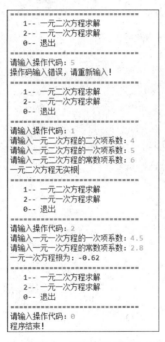

图 5-5 实验运行效果

其中，当求解一元二次方程的根又分为图 5-6 所示的几种情况。

```
请输入一元二次方程的二次项系数：3
请输入一元二次方程的一次项系数：2
请输入一元二次方程的常数项系数：1
一元二次方程无实根
```

```
请输入一元二次方程的二次项系数：1
请输入一元二次方程的一次项系数：4
请输入一元二次方程的常数项系数：2
一元二次方程有两个根为：-0.59、-3.41
```

```
请输入一元二次方程的二次项系数：1
请输入一元二次方程的一次项系数：4
请输入一元二次方程的常数项系数：4
一元二次方程只有一个根为：-2.00
```

```
请输入一元二次方程的二次项系数：0
请输入一元二次方程的一次项系数：5
请输入一元二次方程的常数项系数：9
不是一元二次方程
```

图 5-6 一元二次方程求解

【实验指导】

一元二次方程：
$$ax^2 + bx + c = 0$$
$$s = b^2 - 4ac$$

(1) 当 $a=0$ 时，不是一元二次方程。

(2) 当 $s<0$ 时，一元二次方程无实根。

(3) 当 $s=0$ 时，一元二次方程只有一个根：$-b/2a$。

(4) 当 $s>0$ 时，一元二次方程有两个根：$(-b+s)/2a$ 和 $(-b-s)/2a$。

【程序模板】

```java
package gdlgxy.shiyan;
import java.util.Scanner;
public class SquareEquation {
    /* 功能描述:求方程的根 */
    public static void main(String[] args) {
        //TODO Auto-generated method stub
        Scanner sc = new Scanner(System.in);
        int op;
        do {
            [代码1];//调用方法显示菜单
            System.out.print("请输入操作代码:");
            op = sc.nextInt();
            switch (op) {
                case 0:
                    System.out.println("程序结束!");
                    break;
                case 1:
                    [代码2]//分别输入一元二次方程的系数,输出其根
                    break;
                case 2:
                    [代码3]//分别输入一元一次方程的系数,输出其根
                    break;
                default:
                    System.out.println("操作码输入错误,请重新输入!");
            }
        } while (op != 0);
        sc.close();
    }
    public static void getRoot(double a, double b, double c) {
        double r, r1, r2;
        if (a == 0)
            System.out.println("不是一元二次方程");
        else {
            double s = b * b - 4 * a * c;
```

```
            [代码 4]//分为 s=0,s<0,s>0 三种情况输出其根
        }
    }
    public static void getRoot(double a, double b) {
        double r;
        if (a ==0)
            System.out.println("不是一元一次方程");
        else {
            [代码 5]//求其根
            System.out.printf("一元一次方程根为:%.2f\n", r);
        }
    }
    public static void menu() {              //系统菜单
        System.out.println("==============================");
        System.out.println("        1----元二次方程求解       ");
        System.out.println("        2----元一次方程求解       ");
        System.out.println("        0--退出                    ");
        System.out.println("==============================");
    }
}
```

思考：在一个类中定义两个求根的方法，在另一个类中如何调用？

5.4 拓展练习

【基础知识练习】

一、选择题

1. 下列方法定义中,正确的是(   )。
   A. int method1(int x){return x++;}
   B. double method2(double x) {int a＝1,b＝2;int w＝a＋b;}
   C. void method3(int x){return 0;}
   D. 以上都是

2. 下列方法定义中,方法声明正确的是(   )。
   A. public static max(int a, int b){}   B. public void max(int a, int b){}
   C. public max(int a, int b){}   D. public void max(int a, b, c){}

3. 与 void show(int a,char b,double c){}构成重载的是(   )。
   A. void show(int x,char y,double z){}
   B. int show(int a,double c,char b){}
   C. double show(int x,char y,double z){}
   D. void shows(){}

4. 下列关于方法说法正确的是(   )。
   A. 一个方法内定义的变量可以在方法外使用

B. 一个方法可以有多个返回值

C. 一个方法可以没有返回值

D. 重载的方法可以通过它们的返回值类型的不同来区分

二、填空题

1. 一个方法没有具体返回值的情况，返回值类型用关键字_____表示。

2. 同一个类中多个方法具有相同的方法名，不同的_____称为方法的重载。

3. 下列程序输出结果是_____。

```java
public class Test1 {
    public void printA(){
        System.out.println("方法测试 A");
    }
    public static void main(String[] args) {
        printA();
    }
}
```

【上机练习】

倒序输出一个正整数，例如给出正整数 $n=12345$，希望以各个位数的逆序形式输出，即输出 54321。

提示：

(1) 递归思想。首先输出这个数的个位数，然后再输出前面数字的个位数，直到之前没数字。

(2) 分析数列的递归表达式：

$$f(n)=\begin{cases}\text{print}(n\%10), & n<10 \\ \text{print}(n\%10); f(n/10), & n\geqslant 10\end{cases}$$

第 6 章

Java 数组

6.1 知识提炼

6.1.1 数组概述

数组(array)是一种用一个名字来标识一组有序且类型相同的数据组成的派生数据类型,它占有一片连续的内存空间。主要包括以下几个方面。
- 名字:用以对数组各元素的整体标识,这个名字称为数组名。
- 类型:数组各元素的类型。
- 维数:标识数组元素所需的下标个数。
- 大小:可容纳的数组元素个数(注意,不是字节数)。

数组的主要特性如下。
- 数组可以看成是多个相同数据类型的数据的组合,是对这些相同数据的统一管理。
- 数组变量属引用类型,数组也可以看成是对象,数组中的每个元素相当于该对象的成员变量。
- 数组中的元素可以是任何类型,包括基本类型和引用类型。

6.1.2 一维数组

一维数组中的各个元素排成一行,通过数组名和一个下标就能访问一维数组中的元素。由于数组的所有元素都具有相同的数据类型,因此采用元素数据类型后跟方括号的形式表示数据类型。

1. 一维数组的声明

声明语法如下:

```
数组元素类型 数组名[];
数组元素类型[] 数组名;
```

2. 静态初始化

在定义数组的同时就为数组元素分配空间并赋值。例如:

```
int arr[]={1,2,3};
double decArr[]={1.1,2.2,3.3};
String strArr[]={"Java","BASIC","FORTRAN"};
```

3. 动态初始化

动态初始化是数组声明且为数组元素分配空间,与赋值的操作分开进行。例如:

```
String stringArray =new String[3];      //声明数组包含3个元素
stringArray[0]=new String("How");       //为第一个数组元素开辟空间
stringArray[1]=new String("are");       //为第二个数组元素开辟空间
stringArray[2]=new String("you");       //为第三个数组元素开辟空间
```

4. 默认初始化

数组是引用类型,它的元素相当于类的成员变量,因此数组一经分配空间,其中的每个元素也被按照成员变量以同样的方式隐式初始化。例如:

```
int a[]=new int[5];
System.out.println(a[3]);               //a[3]的默认值为0
```

5. 一维数组的引用

一维数组元素的引用方式如下:

```
数组名[下标]
```

说明:下标可以为整型常数或表达式,下标从0开始。每个数组都有一个属性length指明它的长度,例如,intArray.length指明数组intArray的长度。

6.1.3 二维数组

Java 也支持多维数组。在 Java 语言中,多维数组被看作数组的数组。例如二维数组为一个特殊的一维数组,其每个元素又是一个一维数组。使用二维数组可以方便地处理表格形式的数据。

1. 二维数组的声明

二维数组的声明方式和一维数组的声明方式类似,声明二维数组的一般语法格式如下:

```
类型 数组名[][];
类型[][]数组名;
类型[] 数组名[];
```

例如:

```
int arr[][];
int[][] arr;
int[] arr[];
```

2. 二维数组的初始化

方式一:

```
数据类型 数组名[] [] =new 数据类型[行的个数][列的个数];
```

例如：

```
int[][] arr = new int[3][2];
```

说明：定义了名称为 arr 的二维数组，二维数组中有 3 个一维数组，每一个一维数组中有 2 个元素，一维数组的名称分别为 arr[0]、arr[1]、arr[2]。

方式二：

```
数据类型 数组名[][] = new 数据类型[行的个数][];
```

例如：

```
int[][] arr = new int[3][];
```

说明：二维数组中有 3 个一维数组，每个一维数组都是默认初始化值 null（注意：区别于格式 1），可以对三个一维数组分别分配存储空间，长度可不同。

```
数据类型 数组名[][]={{第 0 行初始值},{第 1 行初始值},...,{第 n 行初始值}};
```

例如：

```
int[][] arr ={{3,8,2},{2,7},{9,0,1,6}};
```

定义一个名称为 arr 的二维数组，二维数组中有三个一维数组，每一个一维数组中具体元素也都已初始化。

- 第一个一维数组 arr[0]={3,8,2};
- 第二个一维数组 arr[1]={2,7};
- 第三个一维数组 arr[2]={9,0,1,6};

3. 二维数组的引用

二维数组的引用表示形式如下：

```
数组名[行号][列号]
```

二维数组是元素为一维数组的数组，其中，各元素数组的长度不尽相同，例如组成二维数组的各个一维数组长度可以不同。共同点是每维数组的索引均从 0 开始。

例如：

```
num[1][0];
```

例如，建立了一个长度为 5 的一维数组，一个长度为 12 的二维数组，源代码如下：

```
public class Demo2_3 {
    public static void main(String args[]) {
        int a[]=new int[5];
        int arr1[][]=new int[3][4];
        a[0]=10;
        a[1]=10+a[0];
```

```
            a[2]=30;
            a[3]=40;
            a[4]=a[1]+a[2];
            arr1[0][0]=0; arr1[0][1]=1; arr1[0][2]=2;
            arr1[1][0]=3; arr1[1][1]=4; arr1[1][2]=5;
            arr1[2][0]=6; arr1[2][1]=7; arr1[2][2]=8;
            System.out.println("a["+0+"] ="+a[0]);
            System.out.println("a["+1+"] ="+a[1]);
            System.out.println("a["+2+"] ="+a[2]);
            System.out.println("a["+3+"] ="+a[3]);
            System.out.println("a["+4+"] ="+a[4]);
            System.out.println("arr1("+0+","+0+") ="+arr1[0][0]);
            System.out.println("arr1("+0+","+1+") ="+arr1[0][1]);
            System.out.println("arr1("+0+","+2+") ="+arr1[0][2]);
            System.out.println("arr1("+1+","+0+") ="+arr1[1][0]);
            System.out.println("arr1("+1+","+1+") ="+arr1[1][1]);
            System.out.println("arr1("+1+","+2+") ="+arr1[1][2]);
    }
}
```

输出结果如下：

```
a[0]=10
a[1]=20
a[2]=30
a[3]=40
a[4]=50
arr1(0,0)=0
arr1(0,1)=1
arr1(0,2)=2
arr1(1,0)=3
arr1(1,1)=4
arr1(1,2)=5
```

6.1.4 数组的空间开辟

数组是一种引用类型，采用顺序存储结构，占用内存中一片连续的存储空间。数组名为该数组对应的内存空间的首地址。

代码如下：

```
int data[] =null;
data =new int[3];                        //开辟一个长度为 3 的数组
data[0] =10;
data[1] =20;
data[2] =30;
```

其空间分配过程如图 6-1 所示。

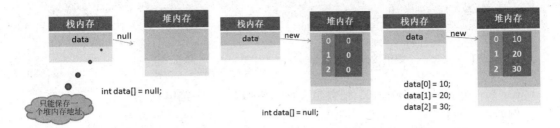

图 6-1 数组内存空间分配过程(1)

如下代码将一个数组赋值给另一个数组,这是地址的传递过程,两个数组指向同一堆内存空间。

```
temp = data;   //int temp[] = data;
temp[0] = 99;
for(int i = 0; i < temp.length; i++) {
    System.out.println(data[i]);
}
```

其空间分配过程如图 6-2 所示。

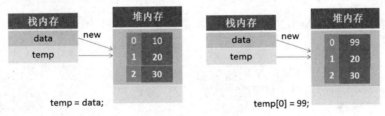

图 6-2 数组内存空间分配过程(2)

6.1.5 数组工具类 Arrays 类

Arrays 类是 JDK 提供的专门用于操作数组的工具类,位于 java.util 包中。用 Arrays 类的方法操作数组,无须自己编码。Arrays 类的常用方法如下。

(1) boolean equals(array1,array2):比较两个数组是否相等。

例如:

```
String[] str1={"1","2","3"};
String[] str2={"1","2",new String("3")};
System.out.println(Arrays.equals(str1, str2));  //结果是 true
```

(2) void sort(array):对数组 array 的元素进行升序排列。

例如:

```java
//给一个数组进行排序
int[] score ={79,65,93,64,88};
Arrays.sort(score);                          //给数组排序
//sort:作用是把一个数组按照由小到大的顺序进行排序
```

(3) String toString(array)：把数组 array 转换成一个字符串。

例如：

```java
int[] a={1,2,3};
System.out.println(a);                       //打印出的是 hashcode 码
System.out.println(Arrays.toString(a));      //打印出的是数组元素
```

(4) void fill(array,val)：把数组 array 所有元素都赋值为 val。

例如：

```java
//fill 方法:把数组中的所有元素替换成一个值
int[] num={1,2,3};
//参数 1:数组对象,参数 2:替换的值
Arrays.fill(num, 6);
System.out.println(Arrays.toString(num)); //打印结果:[6, 6, 6]
```

(5) int binarySearch(array,val)：查询元素值 val 在数组 array 中的下标。

例如：

```java
//binarySearch:通过二分法的方式找对应元素的下标
//使用前提:必须经过排序才可以使用
char[] a={'a','b','c','d','e'};
int i =Arrays.binarySearch(a, 'd');
System.out.println(i);                       //结果是 3

char[] b={'e','a','c','b','d'};
Arrays.sort(b);
int j=Arrays.binarySearch(b, 'e');
System.out.println(j);
```

(6) copyOf(array,length)：把数组 array 复制成一个长度为 length 的新数组。

例如：

```java
//copyOf:把一个原有的数组内容复制到一个新数组中
int[] a={1,2,3};
//参数 1:原数组,参数 2:新数组的长度
int[] b=Arrays.copyOf(a, a.length);
System.out.println(Arrays.toString(b));
//a 和 b 的地址码不同
```

6.2 实例解析

6.2.1 实例1：学生成绩等级判断

【实例要求】

从键盘输入学生成绩到数组中，找出最高分，并输出学生成绩等级。

成绩≥最高分－10　　等级为'A'。
成绩≥最高分－20　　等级为'B'。
成绩≥最高分－30　　等级为'C'。
其余　　　　　　　　等级为'D'。

运行结果如图6-3所示。

```
请输入学生人数：10
请输入10个成绩：
58
96
74
66
42
38
97
58
67
36
最高分：97
学生 0 的成绩为 58，其等级是 D
学生 1 的成绩为 96，其等级是 A
学生 2 的成绩为 74，其等级是 C
学生 3 的成绩为 66，其等级是 D
学生 4 的成绩为 42，其等级是 D
学生 5 的成绩为 38，其等级是 D
学生 6 的成绩为 97，其等级是 A
学生 7 的成绩为 58，其等级是 D
学生 8 的成绩为 67，其等级是 C
学生 9 的成绩为 36，其等级是 D
```

图 6-3　学生成绩等级判断结果

【实例源代码】

新建源文件 ShiyanDemoSum6.java，源代码如下：

```java
package gdlgxy.shiyan;
import java.util.Scanner;
public class ShiyanDemoScore6 {
    /* 功能描述:学生成绩等级判断 */
    public static void main(String[] args) {
        Scanner sc=new Scanner(System.in);
        System.out.print("请输入学生人数:");
        int i=sc.nextInt();
```

```java
        int[] scores=new int[i];
        System.out.print("请输入"+i+"个成绩:");
        for(int j=0;j<i;j++)
            scores[j]=sc.nextInt();
    //求最大值
    int max=scores[0];
    for(int s:scores) {
        if(s>max) max=s;
    }
    System.out.println("最高分:"+max);
    //判断成绩等级
    char grade;
    for(int j=0;j<i;j++){
        if(scores[j]>=max-10) grade='A';
        else if(scores[j]>=max-20) grade='B';
        else if(scores[j]>=max-30) grade='C';
        else grade='D';
        System.out.println("学生"+j+"的成绩为"+scores[j]+",其等级是"+grade);
    }
    }
}
```

【实例解析】

本实例中存储学生成绩的数组长度由用户输入决定,故创建数组时长度不是常量,而是变量。

```
int[] scores=new int[i];
```

求最高分即求数组元素的最大值,而成绩等级的判断要依据实例要求用到分支结构进行运算。

6.2.2 实例2:矩阵转置

【实例要求】

矩阵是排列成若干行若干列的数据表,转置是将数据表的行列互换。即第一行变成第一列、第二行变成第二列、等等。定义一个3×4二维数组,并初始化数组,赋值后,输出二维数组的值,然后求该数组的转置数组,最后输出转置数组的值,运行结果如图6-4 所示。

```
==========原数组的值==========
23    45    67    48
19    49    78    33
26    97    46    29
==========转置数组的值==========
23    19    26
45    49    97
67    78    46
48    33    29
```

图6-4 矩阵转置

【实例源代码】

新建源文件 ShiyanDemoMatrix6.java,源代码如下:

```
package gdlgxy.shiyan;
import java.util.Scanner;
public class ShiyanDemo Matrix6{
    /* 功能描述:矩阵转置算法 */
    public static void main(String[] args) {
        int a[][] = {{23,45,67,48},{19,49,78,33},{26,97,46,29}};
        System.out.print("========原数组的值========");
        for(i =0;i <3;i ++){
            for(int s:a[i]) System.out.print(s +" ");}
        System.out.println();

        int b[][] =new int [4][3];
        int i,j;
        for(i =0;i <3;i ++)
            for(j =0;j <4;j ++)   b[j][i] =a[i][j];
        System.out.print("========转置数组的值========");
        for(i =0;i <4;i ++){
            for(int s:b[i]) System.out.print(s +" ");
        System.out.println();
    }
}
```

【实例解析】

矩阵包含行和列,用二维数组存储矩阵,此实例中用的是二维数组的静态初始化,并用一个二重循环实现转置。则代码如下:

```
for(i =0;i <3;i ++)
    for(j =0;j <4;j ++)   b[j][i] =a[i][j];
```

6.3 上机实验

【实验目的】

- 了解 Java 一维数组、二维数组的声明和定义。
- 掌握 Java 数组的初始化。
- 掌握一维数组元素的访问。
- 掌握二维数组元素的访问。

【实验要求】

在实例 1 的基础上完善学生成绩统计程序,增加排序功能、统计及格率功能。
实验运行效果如图 6-5 所示。

```
请输入学生人数：10
请输入10个成绩：
87
96
54
39
84
56
68
71
92
83
最高分：96
学生 0 的成绩为 87，其等级是 A
学生 1 的成绩为 96，其等级是 A
学生 2 的成绩为 54，其等级是 D
学生 3 的成绩为 39，其等级是 D
学生 4 的成绩为 84，其等级是 B
学生 5 的成绩为 56，其等级是 D
学生 6 的成绩为 68，其等级是 C
学生 7 的成绩为 71，其等级是 C
学生 8 的成绩为 92，其等级是 A
学生 9 的成绩为 83，其等级是 B
学生成绩由低到高为：[39, 54, 56, 68, 71, 83, 84, 87, 92, 96]
及格率为：70.0%
```

图 6-5 实验运行效果

【实验指导】

可用 Arrays 类的 sort()方法对数组进行排序。

另需要增加一个统计方法，统计成绩及格的学生人数，则及格率为及格人数/总人数，这里需要注意的是及格人数和总人数均为整型数据，如果直接除，则结果为 0 或 1。

【程序模板】

```java
package gdlgxy.shiyan;
import java.util.Arrays;
import java.util.Scanner;
public class ScoreCount {
    public static void main(String[] args) {
        Scanner sc=new Scanner(System.in);
        System.out.print("请输入学生人数:");
        int i=sc.nextInt();
        int[] scores=new int[i];
        System.out.print("请输入"+i+"个成绩:");
        for(int j=0;j<i;j++)
            scores[j]=sc.nextInt();
        //求最大值
        int max=scores[0];
        for(int s:scores) {
            if(s>max) max=s;
        }
        System.out.println("最高分:"+max);
        //判断成绩等级
```

```
            char grade;
        for(int j=0;j<i;j++){
            if(scores[j]>=max-10) grade='A';
            else if(scores[j]>=max-20) grade='B';
            else if(scores[j]>=max-30) grade='C';
            else grade='D';
            System.out.println("学生"+j+"的成绩为"+scores[j]+",其等级是"+grade);
        }
        int[] copyScore=Arrays.copyOf(scores, scores.length);
          [代码 1]    //成绩排序
        System.out.println("学生成绩由低到高为:"+[代码 2]);
        int num=[代码 3];
        System.out.printf("及格率为:%.1f%%", [代码 4]);
    }
    public static int count(int[] arr,int x){
        int n=arr.length;
        int num=0;
        for(int i=0;i<n;i++){
            [代码 5]
        }
        return num;
    }
}
```

思考：如何统计各分数段人数？

6.4 拓展练习

【基础知识练习】

一、选择题

1. 下面数组定义正确的有(　　)。
   A. String strs[] = {'a' 'b' 'c'};
   B. String[] strs = {"a", "b", "c"};
   C. String[] strs = new String{"a" "b" "c"};
   D. String[] strs = new String[3]{"a", "b", "c"};

2. 已知数组 arrayInt 由以下语句定义：

```
int[] arrayInt =new int[8];
```

则正确引用数组的最后一个元素的方法是(　　)。
   A. arrayInt[8]   B. arrayInt[0]   C. arrayInt[7]   D. arrayInt[]

3. 定义一维 int 型数组 a[10]后,下面错误的引用是(　　)。
   A. a[1]=1;   B. a[10]=2;   C. a[0]=2*3;   D. a[2]=a[1]*2;

4. 定义了 int 型二维数组 a[6][7]后,数组元素 a[3][4]前的数组元素个数是(　　)。
   A. 24　　　　　B. 25　　　　　C. 18　　　　　D. 17
5. 以下语句 int a[][]={{1,2,3},{4,5,6}};System.out.printf(%d",a[1][1]);的输出结果是(　　)。
   A. 6　　　　　B. 5　　　　　C. 4　　　　　D. 3

二、填空题

1. 当声明一个数组 int arr[]＝new int[5];时,代表这个数组所保存的变量类型是_____,数组名是_____,数组的大小为_____,数组元素下标的使用范围是_____。
2. JVM 将数组存储在_____中。
3. 数组的元素是通过_____来访问,数组 array 的长度为_____,数组最小的下标是_____。
4. 数组下标访问超出索引范围是抛出_____异常。
5. 数组创建后其大小_____改变。
6. 设有数组定义：int MyIntArray[]＝{10,20,30,40,50,60,70};则执行以下几个语句的输出结果是_____。

```
int s=0;
for(int i=0;i<MyIntArray.length;i++)
    if(i%2==1)
        s+=MyIntArray[i];
System.out.println(s);
```

【上机练习】

使用二维数组打印一个 n 行杨辉三角。
```
1
1  1
1  2  1
1  3  3  1
1  4  6  4  1
1  5  10 10 5  1
⋮
```

提示：
(1) 第 1 行有 1 个元素,第 n 行有 n 个元素。
(2) 每一行的第一个元素和最后一个元素都是 1。
(3) 从第 3 行开始,对于非第一个元素和最后一个元素的元素：

```
yanghui[i][j]=yanghui[i-1][j-1] +yanghui[i-1][j];
```

第 7 章

类 和 对 象

7.1 知识提炼

7.1.1 类

类(class)是 Java 语言的最小编译单元,是设计和实现 Java 程序的基础。类是一个抽象概念,是一组具有相同特性(属性)和相同行为(方法)的事物的基础。Java 语言中类的定义包括:成员变量、构造方法、成员方法、内部类、代码块。

```
public class ClassName{
    //成员变量
    //构造方法
    //成员方法
}
```

说明:

(1) 类的访问修饰符可以使用 public 和默认(缺省),但一个.java 文件中只能存在一个 public 修饰的类,且该类名称与文件名称相同。

(2) 类内成员变量一般使用 private 修饰,成员方法一般使用 public 修饰。

7.1.2 构造方法

构造方法(constructor)是创建对象时执行的特殊方法,对实例的成员变量进行初始化,必须和类同名且没有返回值。构造方法的语法格式如下:

```
public class ClassName{
    //成员变量
    public ClassName([参数列表]){
        //成员变量初始化
    }
    //成员方法
}
```

说明:

(1) 构造方法一般使用 public 修饰,方法名称与类名相同。

(2) 构造方法无返回值,不声明返回值类型,也不能使用 void 关键声明。

(3) 当一个类没有声明构造方法时,Java 自动地为该类提供一个无参数的默认构造方法。若声明了构造函数,Java 将不会提供无参数的构造函数。
(4) 构造方法可以被重载。
(5) 构造方法通过 new 运算符调用。

7.1.3 对象的创建与使用

类是对象的模板,对象是类的实例。通过对象使用类,对象的创建使用保留字 new。创建对象的语法格式如下:

```
类名 对象名=new 构造方法([参数])
```

使用对象访问成员变量和成员方法时使用"."运算符。语法格式如下:

```
对象名称.成员变量;对象名称.成员方法([参数列表])
```

对象的创建与使用如下:

```java
class ObjectExample{
/* 功能描述:类定义、对象创建的举例 */
    //[代码块 1]
public int a;
    private char c;
    //[代码块 2]
    public ObjectExample(){}
    public ObjectExample(int a1,char c1){
        a=a1;
        c=c1;
    }
    //[代码块 3]
    public void print(){
        System.out.println(a);
        System.out.println(c);
    }
}
public class ObjectExampleTest {
    public static void main(String[] args) {
    //[代码块 4]
        ObjectExample c1=new ObjectExample();    //声明对象,并实例化
        c1.a=100;                                //通过对象名称使用"."运算符访问公有成员变量
        //c1.c='a';                              //通过对象名称在另外一个类中不能访问私有成员变量
        c1.print();                              //通过对象名称使用"."运算符调用公有成员方法
    //[代码块 5]
        ObjectExample c2;                        //声明对象
        c2=new ObjectExample(200,'b');           //实例化对象,使用 new 关键字调用有参构造函数
        c2.print();                              //调用方法
    //[代码块 6]
        new ObjectExample(300,'c').print();      //匿名参数调用方法
    }
}
```

程序运行结果如图 7-1 所示。

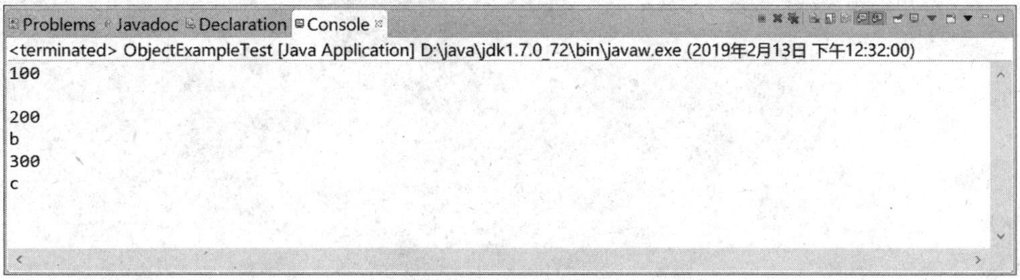
Problems Javadoc Declaration Console
<terminated> ObjectExampleTest [Java Application] D:\java\jdk1.7.0_72\bin\javaw.exe (2019年2月13日 下午12:32:00)
100
200
b
300
c

图 7-1　程序运行结果

说明：

(1) 上述代码保存在名称为 ObjectExampleTest.java 的源文件中。一个.java 文件只能保存在一个 public 类里且名称与文件名称相同。

(2) 文件保存在 ObjectExample、ObjectExampleTest 类里。

(3) 定义 ObjectExample 类。代码块 1 定义了两个成员变量，一个是公有的整型变量 a，一个是私有的字符型变量 c。

(4) 代码块 2 定义了两个构造方法，一个是无参数的构造函数，一个是有参数的构造函数，两个构造函数形成重载。

(5) 代码块 3 定义了一个公有的且无返回值的输出方法 print()，输出 a、b 的值。

(6) 定义 ObjectExampleTest 类，只保存在公有的静态主方法，在主方法中创建并使用对象。

(7) 代码块 4 使用了 ObjectExample 类里无参数构造函数声明并实例化对象。使用对象名称访问公有成员变量，调用公有成员方法。

(8) 代码块 5 首先声明对象，然后调用 ObjectExample 类的有参数构造函数实例化对象。

(9) 代码块 6 使用匿名对象调用公有成员方法。

7.1.4　this 关键字

Java 语言中 this 关键字是指当前对象，可以在类的内部使用 this 关键字访问成员变量、成员方法、构造方法。引用变量的语法格式如下：

```
this.成员变量名
```

调用成员方法的语法格式如下：

```
this.成员方法名([参数列表])
```

调用构造方法的语法格式如下：

```
this([参数列表])
```

例如：

```java
class ThisExample{
    /* 功能描述:this举例 */
    private int a;
    private String s;
    public int getA() {
        return a;
    }
    public void setA(int a) {
        this.a = a;
    }
    public String getS() {
        return s;
    }
    public void setS(String s) {
        this.s = s;
    }
    public ThisExample(int a){
        this.a=a;
    }
    public ThisExample(int a,String s){
        this(a);
        this.s=s;
        this.print();
    }
    public void print(){
        System.out.println(s+a);
    }
}
public class ThisExampleTest {
    public static void main(String[] args) {
        ThisExample te=new ThisExample(2);
        te.setS("这是this关键字的例题1!");
        te.print();
        te=new ThisExample(3,"这是this关键字的例题2!");
    }
}
```

程序运行结果如图 7-2 所示。

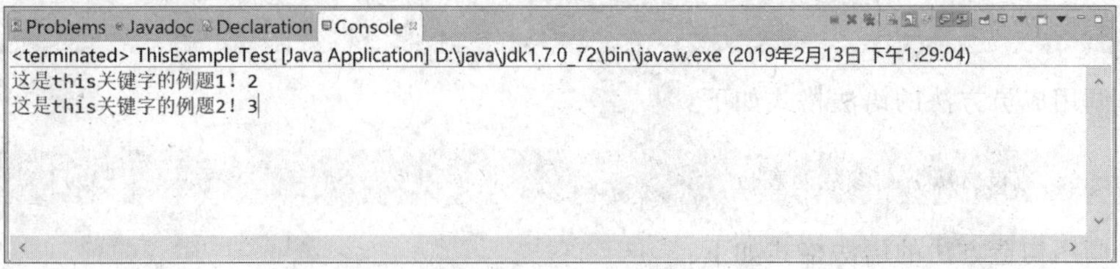

```
这是this关键字的例题1! 2
这是this关键字的例题2! 3
```

图 7-2　程序运行结果

说明：

（1）ThisExample 类中所有变量都有 private，为每个变量编写 getter()和 setter()函数。当局部变量和成员变量同名时，使用 this 关键字来引用成员变量。

（2）构造方法中使用 this 关键字调用其他构造方法时，必须为构造方法的第一条语句。

7.1.5　static 关键字

Java 类中存在两种成员：实例成员和静态成员。使用 static 关键字声明的成员变量称为静态成员变量，又称为类成员。带 static 修饰的属性称为静态属性或类变量；带有 static 修饰的方法称为静态方法。

```java
class StaticExample{
/* 功能描述：static 举例 */
    public static int a;
    public static int b;
    public void print(){
        System.out.println(a+b);
    }
    public static int add(int number1,int number2){
        return number1+number2;
    }
}
public class StaticExampleTest {
    public static void main(String[] args) {
        StaticExample.b=10;
        StaticExample.a=5;
        System.out.println(StaticExample.add(StaticExample.b, StaticExample.a));

        StaticExample se=new StaticExample();
        se.a=20;
        se.b=5;
        System.out.println(se.add(se.a, se.b));
        se.print();

        System.out.println(StaticExample.add(StaticExample.b, StaticExample.a));
    }
}
```

程序运行结果如图 7-3 所示。

```
<terminated> StaticExampleTest [Java Application] D:\java\jdk1.7.0_72\bin\javaw.exe (2019年2月13日 下午4:04:46)
15
25
25
25
```

图 7-3　程序运行结果

说明：

（1）静态方法内可以直接访问静态属性或者调用其他静态方法，不可以访问成员变量，不可以调用非静态成员方法。

（2）静态成员变量分配空间与对象无关，一个类的静态属性成员只有一份。

（3）静态成员变量、静态成员方法可以通过类名访问（在自己类中可以直接访问），也可以通过对象访问。

（4）静态属性是共享的，该类的任何对象均可访问它。

7.2 实例解析

7.2.1 实例1：复数类

【实例要求】

把形如 $z=a+bi(a 、b$ 均为实数)的数称为复数。a 称为实部，b 称为虚部，i 称为虚数单位。两个复数的和依然是复数，它的实部是原来两个复数实部的和，它的虚部是原来两个虚部的和。两个复数的差依然是复数，它的实部是原来两个复数实部的差，它的虚部是原来两个虚部的差。复数的乘法按照以下的法则进行：设 $z_1=a+bi, z_2=c+di(a 、b 、c 、d \in R)$ 是任意两个复数，那么它们的积 $(a+bi)(c+di)=(ac-bd)+(bc+ad)i$。复数除法定义：满足 $(c+di)(x+yi)=(a+bi)$ 的复数 $x+yi(x 、y \in R)$ 叫作复数 $a+bi$ 除以复数 $c+di$ 的商。复数的模是指实部与虚部的平方和的开方。两个复数比较大小通常是比较复数的模。

按照复数的含义及其运算规则设计并编写复数类实现加、减、乘、除、求模运算。加法运算时应满足两个复数相加（类方法和实例方法）、复数与实数相加。

【实例源代码】

新建源文件 ComplexNumber.java，源代码如下：

```
public class ComplexNumber {
    /* 功能描述:复数类,提供加、减、乘、除等功能 */
    private double a;                    //实部
    private double b;                    //虚部
    public double getA() {
        return a;
    }
    public void setA(double a) {
        this.a = a;
    }
    public double getB() {
        return b;
    }
}
```

```java
    public void setB(double b) {
        this.b =b;
    }
    public ComplexNumber() {}                                    //无参构造方法
    public ComplexNumber(double a,double b) {                    //有参构造方法
        this.a=a;
        this.b=b;
    }
    public ComplexNumber add(ComplexNumber c1) {    //两个复数的相加
        ComplexNumber c=new ComplexNumber();
        c.setA(this.a+c1.getA());
        c.setB(this.b+c1.getB());
        return c;
    }
    public ComplexNumber add(double a) {                //复数加实数,方法的重载
        ComplexNumber c=new ComplexNumber();
        c.setA(this.a+a);
        c.setB(this.b);
        return c;
    }
    public static ComplexNumber add(ComplexNumber c1,ComplexNumber c2) {
        //两个复数的相加,但这个是静态方法
        ComplexNumber c=new ComplexNumber();
        c.setA(c1.getA()+c2.getA());
        c.setB(c1.getB()+c2.getB());
        return c;
    }
    public ComplexNumber sub(ComplexNumber c1) {   //减法
        ComplexNumber c=new ComplexNumber();
        c.setA(this.a-c1.getA());
        c.setB(this.b-c1.getB());
        return c;
    }
    public ComplexNumber mult(ComplexNumber c1) {//乘法
        ComplexNumber c=new ComplexNumber();
        c.setA(this.a * c1.getA()-this.b * c1.getB());
        c.setB(this.b+c1.getA()+this.a * c1.getB());
        return c;
    }
    public ComplexNumber div(ComplexNumber c1) {   //除法
        ComplexNumber c=new ComplexNumber();
        double c1a=c1.getA();
        double c1b=c1.getB();
        double temp=c1a * c1a+c1b * c1b;
        c.setA((this.a * c1a+this.b * c1b)/temp);
        c.setB(this.b * c1a-this.a * c1b);
        return c;
    }
```

```java
    public double model() {
        return Math.sqrt(a * a+b * b);
    }
    public boolean isEqual(ComplexNumber c1) {
        if(this.model()==c1.model())
            return true;
        else
            return false;
    }
    public String toString() {
        if(this.a==0&&this.b==0)
            return "0";
        else if(b==0) {
            return String.valueOf(this.a);
        }else if(b<0) {
            return String.valueOf(this.a)+String.valueOf(this.b)+"i";
        }else{
            return String.valueOf(this.a)+"+"+String.valueOf(this.b)+"i";}
    }
}
```

新建源文件 ComplexNumberTest.java，源代码如下：

```java
public class ComplexNumberTest {
    public static void main(String[] args) {
        /* 功能描述:测试 ComplexNumber 类的各个功能 */
        ComplexNumber c1=new ComplexNumber();    //无参构造方法
        c1.setA(10);                              //实部赋值
        c1.setB(5);                               //虚部赋值
        System.out.println(c1.toString());        //输出复数

        ComplexNumber c2=new ComplexNumber(2,10);  //有参构造方法
        System.out.println(c2.toString());         //输出复数

        ComplexNumber c3=c1.add(c2);              //调用 add 方法,计算两个复数的和
        System.out.println(c3.toString());

        ComplexNumber c4=c1.add(10);              //调用 add 方法,计算复数和实数的结果
        System.out.println(c4.toString());

        //使用类的名称调用静态方法计算两个复数的结果并输出
        System.out.println(ComplexNumber.add(c1, c2).toString());

        //上述三个加法操作实现了方法的重载

        //减法
        System.out.println(c1.sub(c1).toString());
        System.out.println(c1.sub(c2).toString());
        System.out.println(c2.sub(c1).toString());
```

```
        //乘法
        System.out.println(c1.mult(c1).toString());
        System.out.println(c1.mult(c2).toString());
        //除法
        System.out.println(c1.div(c1).toString());
        System.out.println(c1.div(c2).toString());
        System.out.println(c2.div(c1).toString());
    }
}
```

程序运行结果如图 7-4 所示。

```
10.00+5.00i
2.00+10.00i
12.00+15.00i
20.00+5.00i
12.00+15.00i
0
8.00-5.00i
-8.00+5.00i
75.00+65.00i
-30.00+107.00i
1.00
0.67-90.00i
0.56+90.00i
```

图 7-4　程序运行结果

【实例解析】

（1）本实例使用了两个源文件，即 ComplexNumber.java 和 ComplexNumberTest.java。ComplexNumber.java 文件实现题目要求的复数类的各个功能，ComplexNumberTest.java 为测试编写的复数类的各个功能。

（2）ComplexNumber 类分为 3 部分，即成员变量、构造方法和成员方法。其中，成员方法又分为类方法和实例方法。

（3）复数存在两个属性，即实部和虚部，实部和虚部都是实数。设置 ComplexNumber 类中两个私有的实数变量 a、b，并且为 a、b 设置 getter()、setter() 方法。

（4）在私有成员变量 a、b 的 setter() 方法中使用了 this 关键字，当局部变量和成员变量同名时且要使用成员变量时，使用 this 关键字应用。

（5）构造方法必须和类的名称相同且无返回值和返回值类型，构造方法可以重载。本题编写的无参数和有参数构造方法有两个，构造方法的作用是初始化对象的属性。如果不编写构造方法，系统会自动添加无参数的构造方法；如果编写了构造方法，系统将不会添加无参数构造方法。编写复数类的构造方法时编写有参数的构造方法 public ComplexNumber(double a, double b)，系统就不会自动添加无参数的构造方法。有时会用到无参数的构造方法，可以自己添加无参数的构造方法 public ComplexNumber()。创建对象时使用 new 关键字调用构造方法。

（6）复数的加法运算规则是实部和虚部对应相加，其结果仍然是复数，所以加法的返回值类型是 ComplexNumber。本实例采用三个方法分别实现复数的加运算，这三个方法名称和返回值类型相同，但参数列表不同，因此形成了重载。这三个方法分别是 public ComplexNumber add(ComplexNumber c1)、public ComplexNumber add(double a)、public static ComplexNumber add(ComplexNumber c1,ComplexNumber c2)。第一个方法和第三个方法的功能相同，都是两个复数相加，但是一个是静态方法，一个是实例方法；第二个方法计算的是一个复数和一个实数相加。

（7）求模运算中使用 Math.sqrt(double a)方法是求平方根的方法。Math 是 java.lang 包下的类，该类包含常用的数学方法，比如：

- Math.sqrt() 计算平方根。
- Math.cbrt() 计算立方根。
- Math.pow(a,b) 计算 a 的 b 次方。
- Math.max(,) 计算最大值。
- Math.min(,) 计算最小值。
- Math.abs() 取绝对值。
- Math.rint() 四舍五入，返回 double 值。注意，如果是 5 就会取偶数。
- Math.round() 四舍五入，float 时返回 int 值，double 时返回 long 值。
- Math.random() 取得一个[0,1)范围内的随机数。

（8）ComplexNumberTest.java 文件创建 ComplexNumberTest 类是 ComplexNumber 的测试类，类里有主函数 public static void main(String[] args)。类是模板，若想使用类，必须进行实例化，通过对象使用属性和方法。

（9）ComplexNumberTest 中存在三种创建对象，即 ComplexNumber c1 = new ComplexNumber()、ComplexNumber c2 = new ComplexNumber(2,10)、ComplexNumber c3=c1.add(c2)。第一种使用 new 创建对象并调用无参数构造方法初始化对象；第二种使用用 new 创建对象并调用有参数的构造方法；第三种加法运算的返回值是复数。

（10）System.out.println(ComplexNumber.add(c1,c2).toString())，通过 ComplexNumber 类调用静态方法计算两个复数的相加结果。该方法的返回值是复数，通过方法返回值调用 toString()方法，使用的是匿名对象调用方法。

（11）方法重载时方法的名称一定要相同，而参数表必须不同。用不同的参数类型或个数来区分不同的方法体。方法的重载和方法的返回类型、修饰符无关。

（12）类名要有意义，最好要见名知意，且首字母应大写。

7.2.2 实例 2：银行账户类

【实例要求】

一个银行账户(Account)要包含银行账号(number)、密码(password)、姓名(name)、账户余额(balance)、开户时间(beginDate)等属性。账户能进行密码加密、存款、取款、重置密码、返回账户信息等功能。账户密码加密的算法较多，任选一种加密方法即可。存款、取款、重置密码功能首先要验证账户和密码是否正确，若正确就继续操作，若不正确则结束操作。当取款时，按照存款是否成功返回存款的状态码。

【实例源代码】

新建源文件 Account.java,源代码如下：

```java
public class Account {
    /* 功能描述:银行账户类,提供密码加密、存款、取款、重置密码、返回账户信息等功能 */
    private String number;                    //银行账号
    private String password;
    private String name;
    private String beginDate;
    private double balance;
    public String getPassWord() {
        return password;
    }
    public void setPassWord(String password) {
        this.password = encrypt(number,password);
    }
    public String getNumber() {
        return number;
    }
    public void setNumber(String number) {
        this.number = number;
    }
    public String getName() {
        return name;
    }
    public void setName(String name) {
        this.name = name;
    }
    public String getBeginDate() {
        return beginDate;
    }
    public void setBeginDate(String l) {
        this.beginDate = l;
    }
    public double getBanlance() {
        return balance;
    }
    public void setBanlance(double balance) {
        this.balance = balance;
    }
    public Account(){}
    public Account(String number, String password, String name, String beginDate, double balance) {
        this.number = number;
        this.password = encrypt(number,password);
```

```java
        this.name =name;
        this.beginDate =beginDate;
        this.balance =balance;
    }
    public static String encrypt(String number, String password) {
                //使用账号名称给密码加密,这只是模拟加密方式,并不是银行实际加密方式
        char[] p =password.toCharArray();//字符串转字符数组
        int n =p.length;                    //密码长度
        char[] c =number.toCharArray();
        int m =c.length;                    //字符串长度
        for (int k =0; k <m; k++) {
            int mima =c[k] +p[k%n];         //加密
            c[k] = (char) mima;
        }
        return new String(c);
    }
    public boolean deposit(String number,String password,double b) {    //存款
        if(number.equals(number)&&this.password.equals(encrypt(number,password))){
                                            //若密码和账号都正确存入
            this.balance+=b;
            return true;
        }else{
            return false;
        }
    }
class State{
    public boolean b;
    public int n;
    public State(boolean b,int n){
        this.b=b;
        this.n=n;
    }
}
public State withdraw(String number,String password,double b){
    if(!number.equals(number))
    {
        return new State(false,1);
    }else if(!this.password.equals(encrypt(number,password)))
    {
        return new State(false,2);
    }
    else if(b>this.balance)
    {
        return new State(false,3);
    }
    else{
        this.balance-=b;
        return new State(true,0);
    }
}
```

```java
    public State reset(String number, String passwordOld, String passwordnew1, String passwordnew2){
        //重置密码
        if(!number.equals(number))
        {
            return new State(false,1);
        }else if(!this.password.equals(encrypt(number,passwordOld)))
        {
            return new State(false,2);
        }
        else if(!passwordnew1.equals(passwordnew2))
        {
            return new State(false,3);
        }
        else{
            this.setPassWord(passwordnew1);
            return new State(true,0);
        }
    }
    public String toString(){
        return this.number+"   "+this.password+"\n"+this.balance+"\n"+this.name+"   "+this.beginDate;
    }
}
```

新建源文件 AccountTest.java,源代码如下:

```java
import java.util.Date;
public class AccountTest {
    /* 功能描述:测试 Account 类的各个功能 */
    public static void main(String[] args) {
        Account a1=new Account();
        a1.setName("张三");
        a1.setNumber("1234567890101213");
        a1.setPassWord("223456");
        a1.setBanlance(10000);
        a1.setBeginDate(new Date().toString());
        System.out.println(a1.toString());
        Account a2= new Account("1234567890101918","123456","李四",new Date().toString(),800);
        System.out.println(a2.toString());
        if(a1.deposit(a1.getNumber(), "223456", 1000))
        {
            System.out.println(a1.toString());
        }else{
            System.out.println("密码有误!");
        }
```

```
            int sn=a2.withdraw("12345678890101918", "123456", 1000).n;
            if(sn==0){
                System.out.println(a2.toString());
            }else if(sn==1){
                System.out.println("账号有误!");
            }else if(sn==2){
                System.out.println("密码有误!");
            }else if(sn==3){
                System.out.println("余额不足!");
            }
            sn=a2.reset("12345678890101918", "123456", "321465", "321465").n;
            if(sn==0){
                System.out.println(a2.toString());
            }else if(sn==1){
                System.out.println("账号有误!");
            }else if(sn==2){
                System.out.println("密码有误!");
            }else if(sn==3){
                System.out.println("两次新密码不同!");
            }
        }
    }
```

程序运行结果如图 7-5 所示。

```
1234567890101213    cdfhjlijldffcddg
10000.0
张三  Wed Feb 13 17:47:27 CST 2019
1234567890101918    bdfhjlhjldffbkdl
800.0
李四  Wed Feb 13 17:47:27 CST 2019
1234567890101213    cdfhjlijldffcddg
11000.0
张三  Wed Feb 13 17:47:27 CST 2019
余额不足!
1234567890101918    dddhkkjjjdgedkbl
800.0
李四  Wed Feb 13 17:47:27 CST 2019
```

图 7-5　程序运行结果

【实例解析】

(1) 按照要求创建 Account 类，在这个类中包含 private String number、private String password、private String name、private String beginDate、private double balance 私有属性。在 Java 设计类时，成员属性通常使用私有的，同时保持类的封装性。为了在类外使用这些属性，通常为每个类添加 getter() 和 setter() 方法。从类外能受控地访问类属性。

在编译器中可以快速生产 getter() 和 setter() 方法。

书写上类内包含的私有成员变量：Source→Generate Getters and Setters。弹出的对话框如图 7-6 所示。

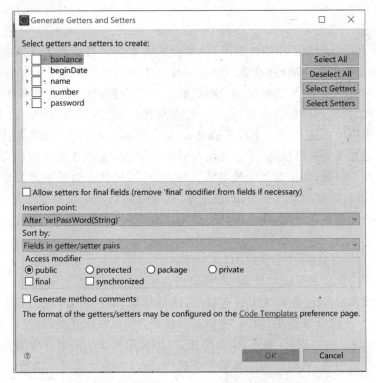

图 7-6　getter()和 setter()方法的设置对话框

(2) 在 Java 中，可以将一个类定义在另一个类里或者一个方法里，这样的类称为内部类。广泛意义上的内部类一般包括四种：成员内部类、局部内部类、匿名内部类和静态内部类。成员内部类是最普通的内部类，它的定义位于另一个类的内部。本实例中存在成员内部类 class State{}。

(3) 在 AccountTest.java 源文件中使用 java.util.Date 包。Date 类是 Java 提供的一种时间操纵类。本实例中使用该类中的一个方法获取当前时间并转化成字符串 new Date().toString()。

7.3　上机实验

【实验目的】

- 了解类的设计。
- 掌握类的定义方法。
- 掌握成员变量、成员方法、构造方法的定义。
- 掌握 this、static 关键字的使用。
- 掌握类封装思想及方法。
- 熟悉类和对象的关系。

【实验要求】

使用Java语言分析以下功能性要求,抽象出核心类并通过编程实现该类,设计一个测试类,测试该核心类的各个属性、功能。学校要建立图书馆座位管理系统有的要求如下。

每个座位都有一个编号、座位状态、座位描述、是否损害、所属机构等属性。每个座位的编号含有具体的含义,前两位表示区域,3、4位表示楼层,剩余几位表示座位编号。可以根据座位编号得出座位描述。每一个没有损害的座位都可以被申请、暂时离开、释放三种情况。每位学生通过学号操纵座位,学生使用学号申请座位,座位会记录使用者的学号,把座位状态更改为使用状态;学生通过学号申请暂时离开,座位状态会被更改为暂离状态;当学生不再使用座位时,需要释放座位资源,清空学号,将座位状态更改为空闲。在使用状态、暂离状态的座位不会被再次分配,在空闲状态的座位会参与分配。

【实验指导】

分析图书馆座位管理系统功能,要求抽象出该系统的核心类是座位类(Seat)。通过上述描述可以得出座位类包含属性应该有编号、座位状态、座位位置描述、是否损坏、所属机构、使用者学号。座位存在使用、暂离、空闲三种状态,所以在设计类时,把该属性定义为整型(int或byte或short)。座位编号、座位描述、使用者学号应使用字符串类型(String)。该系统部署在一个机构内,所有的座位都属于该机构,所以在设计类时可以把此属性设置成为静态成员变量。

座位类除了包括私有属性的getter()和setter()方法及构造方法外,还应该包括座位申请、座位释放、座位保留、获得座位信息的核心方法。

【程序模板】

```java
public class Seat {
    /* 功能描述:座位类包含属性应该有编号、座位状态、座位位置描述、是否损坏、所属机构、使用者学号;还包括座位申请、座位释放、座位保留、获得座位信息的核心方法 */

    private static String institution="广东理工学院图书馆";
        //座位编号
    private String ID;
        //座位状态 1表示空闲,2表示使用,3表示暂离
    private int status;
        //座位描述
    private String description;
        //使用者ID
    private String userID;
        //是否损坏
    private boolean normal;

    public String getID() {
        return ID;
    }
```

```
public void setID(String iD) {
    ID = iD;
}
public int getStatus() {
    return status;
}
public void setStatus(int status) {
    this.status = status;
}
public String getDescription() {
    return description;
}
public void setDescription(String description) {
    this.description = description;
}
public String getUserID() {
    return userID;
}
public void setUserID(String userID) {
    [代码 1]
}
public boolean isNormal() {
    return normal;
}
public void setNormal(boolean normal) {
    this.normal = normal;
}
public Seat(){}                                              //无参数的构造函数
public Seat(String ID,int status,boolean normal){            //带参数的构造函数
    [代码 2]
}
public boolean applySeat(String userID) {        //返回值为 true 表示申请成功,
                                                 返回值为 false 表示申请失败
    if(normal==true&&status!=2&&status!=3){      //若座位正常且无人使用则可分
                                                 配,否则不予分配
        [代码 3]
    }else{
        [代码 4]
    }
}
public boolean releaseSeat(String userID){       //释放座位
    if(userID.equals(this.userID)){              //判断座位是否属于该人,若是,则释放座位
        [代码 5]

    }else{
        return false;
    }
}
```

```java
    public boolean reservedSeat(String userID){    //使用人员暂时离开
        if(userID.equals(this.userID)){
            this.status=3;
            return true;
        }else{
            return false;
        }
    }
    public String introduction(){                   //返回座位信息
        String temp=institution;
        switch(ID.substring(0, 2)){
            case "01":temp+="高要校区";break;
            case "02":temp+="鼎湖校区";break;
        }
        temp= temp+ ID.substring(2, 4)+"楼"+ ID.substring(4, ID.length())+"号座位。";
        this.description=temp;
        if(userID!=null){
            this.description=this.description+"使用者学号"+userID;
        }
        return this.description;
    }
}
```

新建源文件 SeatTest.java，源代码如下：

```java
import java.util.Scanner;
public class SeatTest {
    /* 功能描述:座位类的功能测试 */
    public static void main(String[] args) {
        Seat s1=new Seat();                  //通过 new 调用 Seat 类无参构造函数创建座位对象
        s1.setID("0102024");                 //设置位置编号
        s1.setNormal(true);                  //设置座位的可用性
        s1.setStatus(1);                     //设置座位的状态
        if(s1.applySeat("20192104100")){
            System.out.println(s1.introduction()+"请使用");
        }//使用学号申请账号,若成功,则输出座位信息
        if(s1.reservedSeat("20192104100")){
            System.out.println(s1.introduction()+"为您保留");
        }
        if(s1.releaseSeat("20192104100")){
            System.out.println(s1.introduction()+"已经释放");
        }
        Seat[] sarry=new Seat[4];       //创建 4 个座位数组
        for(int i=0;i<sarry.length;i++){
            [代码 6]//调用有参构造函数实例化对象
        }
```

```
        Scanner sc=new Scanner(System.in);    //使用输入时,需要导入
                                                java.util.Scanner包
        System.out.println("请你输入你的学号!");
        String userID=sc.nextLine();
        if(sarry[0].applySeat(userID)){
            System.out.println(sarry[0].introduction());;
        }
        sc.close();
    }
}
```

思考：如何添加座位报修功能？

7.4 拓展练习

【基础知识练习】

一、选择题

1. 方法的形参(　　)。

　　A. 可以没有　　　　　　　　B. 至少有一个

　　C. 必须定义多个形参　　　　D. 只能是简单变量

2. 下列(　　)类声明是正确的。

　　A. public void A1{...}　　　　B. public class Move(){...}

　　C. public class void number{...}　　D. public class Car{...}

3. 下面对构造方法的描述不正确的是(　　)。

　　A. 构造方法名与类名相同

　　B. 构造方法不显式声明返回类型,可以使用 void 声明

　　C. 构造方法可以重载

　　D. 构造方法可以设置参数

4. 设 Circle 为已定义的类名,下列声明 A 类的对象 c1 的语句中正确的是(　　)。

　　A. float Circle c1;　　　　　B. public Circle c1＝Circle();

　　C. Circle c1＝new int();　　　D. Circle c1＝new Circle();

5. 有一个类 A,以下为其构造方法的声明,其中正确的是(　　)。

　　A. void A(int x){...}　　　　B. public A(int x){...}

　　C. public a(int x){...}　　　D. static A(int x){...}

二、填空题

1. 面向对象的特征是_____、_____、_____。

2. 在 Java 程序中定义的类的成员是 _____、_____。

3. _____运算符的作用是根据对象的类型分配内存空间。当对象拥有内存空间时,会自动调用类中的_____。

4. Java 中通常将属性声明为私有的_____,再通过公共的_____ setter()和 getter()

方法设置和获取,实现对属性的操作。

5. 阅读下列程序,程序运行的结果是_____。

```java
public class StaticDemo {
    public static void main(String[] args) {
        System.out.print(Static.a+" ");
        Static s1=new Static(1,2);
        System.out.print(s1.a+" ");
        s1.a=11;
        System.out.print(Static.a);
    }
}
class Static{
    public static int a=10;
    public int b;
    public Static(int a,int b){
        this.a=a;
        this.b=b;
    }
}
```

【上机练习】

开发一个简单的购书系统,输出图书信息,包括编号、书名、单价、库存。顾客要购书时输入图书编号,并根据提示输入购买数量。顾客购书完毕,输出顾客的订单信息,包括订单号、订单明细、订单额。程序运行界面如图 7-7 所示。

练习分析:通过要求可知,简单的购书系统存在 3 个类:图书类(Book)包含的属性有(图书编号 id、图书名称 name、图书单价 price、库存数量 storage);订单项(OrderItem)包含属性(图书 book、购买数量 num);订单类(Order)包含的属性有(订单号 orderID、订单总额 total、订单列表 items、添加订单项方法、计算总价格),在测试类中模拟形成简单图书购买系统。

代码模板是 Book.java,源代码如下:

```java
public class Book {
    /* 功能描述:
     */
}

OrderItem.java
public class OrderItem {
    /* 功能描述:
     */
}

Order.java
```

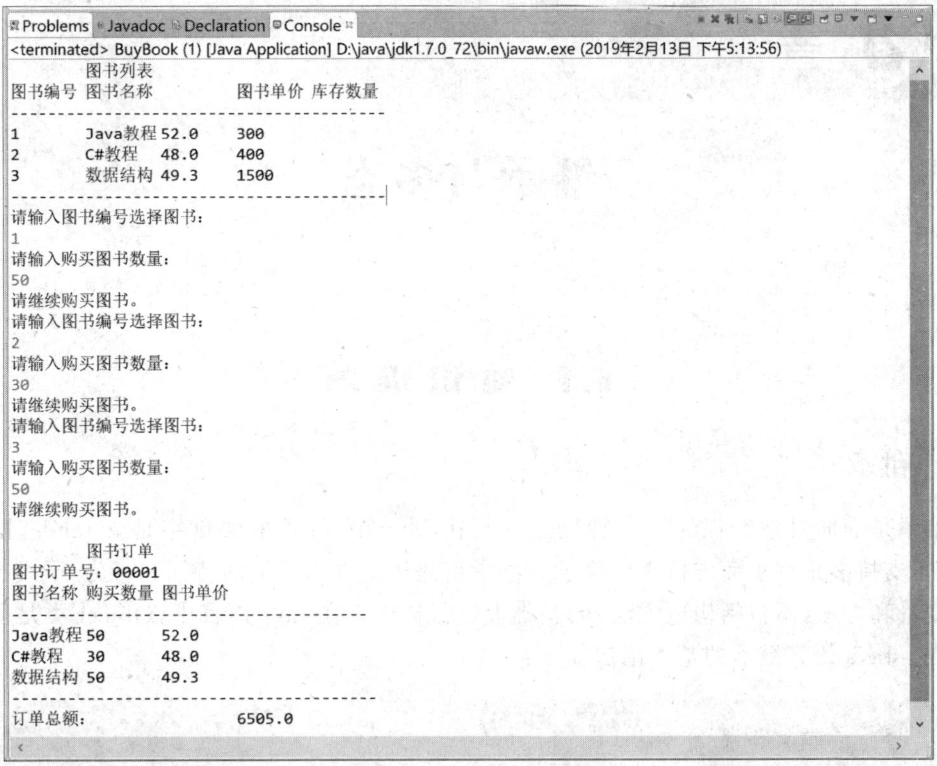

```
图书列表
图书编号 图书名称        图书单价 库存数量
----------------------------------------
1       Java教程 52.0    300
2       C#教程   48.0    400
3       数据结构 49.3    1500
----------------------------------------
请输入图书编号选择图书:
1
请输入购买图书数量:
50
请继续购买图书。
请输入图书编号选择图书:
2
请输入购买图书数量:
30
请继续购买图书。
请输入图书编号选择图书:
3
请输入购买图书数量:
50
请继续购买图书。

        图书订单
图书订单号: 00001
图书名称 购买数量 图书单价
----------------------------------------
Java教程 50       52.0
C#教程   30       48.0
数据结构 50       49.3
----------------------------------------
订单总额:                6505.0
```

图 7-7 程序运行结果

```java
public class Order {
    /* 功能描述:
     */
}
```

第 8 章

继承与多态

8.1 知识提炼

8.1.1 继承

继承是面向对象编程的三大特征之一。在 Java 中,继承的关键字是 extends,Java 语言,它不支持多重继承而支持多层继承。继承就是子类继承父类的特征和行为,子类在拥有父类变量和方法(不包括构造方法)的基础上进行扩展。在 Java 语言中,Object 类是所有类的祖先。Java 语言继承的基本语法如下:

```
class SupClassName{
    //变量
    //构造方法
    //方法
}
class ClassName extends SupClassName{
    //变量
    //构造方法
    //方法
}
```

说明:
(1) 一个子类只能有一个父类,一个父类可以有多个子类。
(2) 子类继承父类的方法和属性,但是不继承父类的构造方法。
(3) 父类不拥有子类新增加的属性和方法。
(4) 子类对父类成员的访问权限。
- 子类不能访问父类的私有成员。
- 子类能够访问父类的公有成员和保护成员。
- 子类和父类在同一包中可以访问默认权限成员。

8.1.2 子类构造方法

在 Java 语言中,子类不继承父类的构造方法。子类的构造方法对所有成员变量进行初始化,子类通过 super 关键字调用父类的构造方法。若子类的构造方法不调用父类的构造方法,则 Java 会默认调用父类无参数的构造方法。若父类不存在无参数的构造方法,就会报错。

```java
class TestA{
    public int a;
    public TestA(){
        System.out.println("TestA 的无参数构造方法!");
    }
    public TestA(int a){
        this.a=a;
        System.out.println("TestA 的有参数构造函数,参数 a="+a);
    }
}
class TestB extends TestA{
    public int b;
    public TestB(){
        System.out.println("TestB 的无参数构造方法!");
    }
    public TestB(int a,int b){
        super(a);
        System.out.println("TestB 的有参数构造方法,b="+b);
    }
}
class TestC extends TestB{
    public int c;
    public TestC(){
        System.out.println("TestC 的无参数构造方法!");
    }
    public TestC(int a,int b,int c){
        super(a,b);
        System.out.println("TestC 的无参数构造方法,c="+c);
        this.c=c;
    }
}
public class ConstruTest {
    public static void main(String[] args) {
        TestA ta1=new TestA();
        System.out.println("-------------------------");
        TestA ta2=new TestA(1);
        System.out.println("-------------------------");
        TestB tb1=new TestB();
        System.out.println("-------------------------");
        TestB tb2=new TestB(2,2);
        System.out.println("-------------------------");
        TestC tc1=new TestC();
        System.out.println("-------------------------");
        TestC tc2=new TestC(3,3,3);
    }
}
```

程序运行结果如图 8-1 所示。

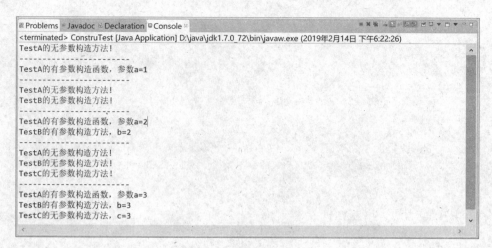

图 8-1　程序运行结果

说明：

（1）使用 super([参数列表])调用父类的构造方法，必须放在子类构造方法的第一句。

（2）当一个子类没有声明构造方法时，Java 为该类提供默认构造方法。该默认构造方法会默认调用 super()执行无参数构造方法。若子类构造方法没有调用父类构造方法并且也没有用 this 关键字调用其他构造方法时，java 将默认执行父类的无参数构造方法。

（3）创建子类对象时构造方法的执行顺序是先执行子类构造方法，然后逐级执行父类构造方法。

8.1.3　多态

多态意为一个名字可以具有多种语义。在面向对象程序设计中，多态性主要表现为类声明的变量可以指向多种不同的对象，具有多种对象的能力。在 Java 中，对象的变量可以引用本类对象，也可以引用子类对象，这称为对象变量多态性。同一方法调用形式实际能调用不同版本的方法的现象称为方法多态性。

```java
class Parent{
    int x=100;
    public void print(){
        System.out.println("这是parent的输出方法,x="+x);
    }
}
public class Child extends Parent {
    int x=200;
    public void print(){
        System.out.println("这是Child的输出方法,x="+x);
    }
    public static void main(String[] args) {
        Parent p1=new Parent();
        Parent p2=new Child();
        p1.print();
```

```
        p1=p2;
        p1.print();
    }
}
```

程序运行结果如图 8-2 所示。

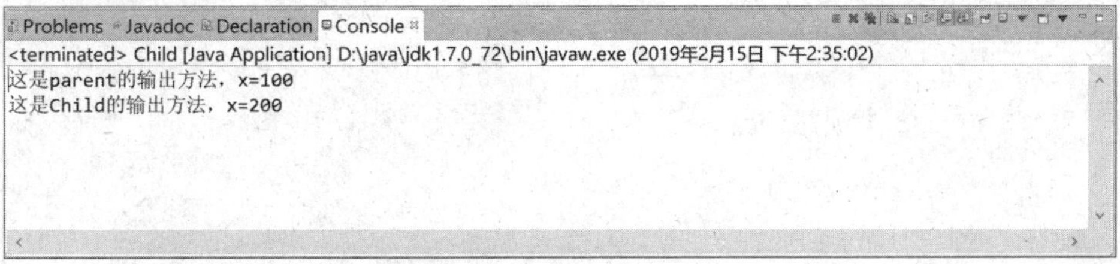

```
Problems  Javadoc  Declaration  Console
<terminated> Child [Java Application] D:\java\jdk1.7.0_72\bin\javaw.exe (2019年2月15日 下午2:35:02)
这是parent的输出方法, x=100
这是Child的输出方法, x=200
```

图 8-2　程序运行结果

说明：

（1）当子类变量名称和父类变量名称相同时，子类变量会覆盖父类成员变量。若想在子类访问被子类覆盖的成员变量，可以使用 super 关键字。

（2）子类可以重写父类成员方法。当子类对象被创建时，调用子类重写的成员方法。使用 final 关键字修饰的方法不可被重写，使用 final 关键字修饰的类不能被继承。

（3）声明对象时，其指定的类型并不是对象真正的类型，对象的真正类型是由创建对象时调用的构造方法决定的，这是对象变量的多态。

（4）使用父类声明的对象引用实例对象后，调用父类存在的方法会调用子类重写的方法。实际调用那个方法时由运行时的动态绑定决定，而不是由声明对象变量的类型决定。

（5）父类对象引用子类实例，不能访问或调用父类中没有声明的成员变量或方法。

8.2　实例解析

8.2.1　实例 1：银行信用卡类

【实例要求】

银行除了普通用户外，还存在借贷用户，设计一个借贷用户，在具有普通银行账户属性和功能的基础上添加以下属性和方法：信用评分、最大借贷额度、本月当还金额、总应还金额，添加还款功能，重写取款、返回账户信息功能。还款功能是指从当前账户余额中扣除本月应还金额，若账户余额大于本月应还金额时，扣除本月应还金额，本月应还金额置 0，总应还金额扣除本月应还金额。若账户余额不够，返回还款失败、扣除相应的信用评分。取款功能是指账户余额足够扣除应取金额，若金额不够，首先查询信用评分，若信用评分满足，再查询是否超出最大借贷金额，若没有，则让取钱，若超过，则禁止取钱。返回账户信息功能是在原信息的基础上添加借贷账户的信息。

注：在银行普通账户类第 7 章的实例解析中已经给出。

【实例源代码】

新建源文件 CheckingAccount.java，源代码如下：

```java
package account;
public class CheckingAccount extends Account {
  /* 功能描述:信用卡账户类 */
    private int creditRating;                    //信用等级
    private double payments_1;                   //当月应还金额
    private double payments_total;               //总应还金额
    private double max;                          //最大透支金额
    public int getCreditRating() {
        return creditRating;
    }
    public void setCreditRating(int creditRating) {
        this.creditRating =creditRating;
    }
    public double getPayments_1() {
        return payments_1;
    }
    public void setPayments_1(double payments_1) {
        this.payments_1 =payments_1;
    }
    public double getPayments_total() {
        return payments_total;
    }
    public void setPayments_total(double payments_total) {
        this.payments_total =payments_total;
    }
    public double getMax() {
        return max;
    }
    public void setMax(double max) {
        this.max =max;
    }
    public CheckingAccount(){}
    public CheckingAccount(int creditRating, double payments_1, double payments_total, double max) {
        super();
        this.creditRating =creditRating;
        this.payments_1 =payments_1;
        this.payments_total =payments_total;
        this.max =max;
    }
    public CheckingAccount(String number, String password, String name, String beginDate, double banlance) {
        super(number, password, name, beginDate, banlance);
    }
```

```java
    public CheckingAccount(String number, String password, String name, String
beginDate, double banlance, int creditRating, double payments_1, double payments_
total, double max) {
        super(number, password, name, beginDate, banlance);
        this.creditRating=creditRating;
        this.payments_1=payments_1;
        this.payments_total=payments_total;
        this.max=max;
    }
    public State withdraw(String number,String password,double b){
        if(!number.equals(number))
        {
            return new State(false,1);
        }else if(!super.getPassWord().equals(encrypt(number,password)))
        {
            return new State(false,2);
        }
        else if(b>super.getBanlance()+max-payments_total)
        {
            return new State(false,3);
        }
        else{
            if(super.getBanlance()-b<0){
                if(this.creditRating>85){
                    payments_total+=b;
                }else{
                    return new State(true,4);
                }

            }else{
                super.setBanlance(super.getBanlance()-b);
            }
            return new State(true,0);
        }
    }
    public boolean pay(){           //还款
        if(super.getBanlance()-payments_1<0){
            this.creditRating-=10;     //降低信用等级
            return false;

        }else{
            super.setBanlance(super.getBanlance()-payments_1);
            payments_total-=payments_1;
            payments_1=0;
            return true;
        }
    }
    public String toString(){
        return super.toString()+"\n"+"信用评分:"+this.creditRating+" 本月应还金
```

额"+this.payments_1+" 总应还金额:"+this.payments_total+" 剩余可借金额:"+(this.max-this.payments_total);
        }
    }

新建源文件 CheckingAccountTest.java,源代码如下：

```java
package account;
import java.util.Date;
public class CheckingAccountTest {
    /* 功能描述:信用卡账户类测试 */
    public static void main(String[] args) {
        CheckingAccount ca1=new CheckingAccount();
        ca1.setName("张三");
        ca1.setNumber("12345678901101213");
        ca1.setPassWord("223456");
        ca1.setBanlance(1000);
        ca1.setBeginDate(new Date().toString());

        ca1.setCreditRating(80);
        ca1.setMax(10000);
        ca1.setPayments_total(7000);
        ca1.setPayments_1(1000);
        System.out.println("---------个人信息-----------");
        System.out.println(ca1.toString());
        System.out.println("---------还款-----------");
        if(ca1.pay()==true){//还款
            System.out.println("本月已还清!");
        }else{
            System.out.println("账户余额不足,无法还清本月账单!");
        };
        System.out.println("---------还款后个人信息-----------");
        System.out.println(ca1.toString());
        System.out.println("---------取款-----------");
        int sn=ca1.withdraw("12345678901101213", "223456", 1000).n;
        if(sn==0){
            System.out.println("成功取出!"+ca1.toString());
        }else if(sn==1){
            System.out.println("账户有误!");
        }else if(sn==2){
            System.out.println("密码有误!");
        }else if(sn==3){
            System.out.println("已超出最大借贷金额!");
        }else if(sn==4){
            System.out.println("信用等级不足,信用等级为 85 分才可借钱,该账户信用等级为:"+ca1.getCreditRating());
        }
    }
}
```

程序运行结果如图 8-3 所示。

```
Problems  Javadoc  Declaration  Console
<terminated> CheckingAccountTest [Java Application] D:\java\jdk1.7.0_72\bin\javaw.exe (2019年2月15日 下午3:37:04)
----------个人信息----------
账号：1234567890101213 密码：cdfhjlijldffcddg 账户余额：1000.0 姓名：张三 开户时间：Fri Feb 15 15:37:04 CST 2019
信用评分：80 本月应还金额：1000.0 总应还金额：7000.0 剩余可借金额：3000.0
----------还款----------
本月已还清！
----------还款后个人信息----------
账号：1234567890101213 密码：cdfhjlijldffcddg 账户余额：0.0 姓名：张三 开户时间：Fri Feb 15 15:37:04 CST 2019
信用评分：80 本月应还金额：0.0 总应还金额：6000.0 剩余可借金额：4000.0
----------取款----------
信用等级不足，信用等级为85分才可借钱，该账户信用等级为：80
```

图 8-3　程序运行结果

【实例解析】

（1）根据题目要求，继承已经存在的类，若在一个项目但不在一个包内可以导入包，若不在同一个包内可以将文件复制过来。

（2）继承就是继承父类的属性和方法，根据要求添加属性和方法或者重写方法。CheckingAccount 类重写了取款方法、添加了还款方法。取款失败分为四种情况：账户有误、密码有误、超出最大取款金额、信用评级不合格。State 是 Account 中的内部类。

（3）本实例重载了构造方法，一个无参数的，两个有参数的（参数的数量不同）。使用 new 关键字创建构造类实例时，根据构造方法参数列表的不同调用合适的构造方法。

（4）super 关键字用来指代父类对象，一般使用 super 关键字调用被子类覆盖的成员变量和父类构造方法。使用 super 关键字调用父类成员变量时，该变量必须满足访问规则。super 关键字调用父类成员变量时，应放在子类构造方法的第一句。

（5）若子类构造方法不调用父类构造方法或不调用本身构造方法时，Java 会为子类构造添加无参数的父类构造方法。若父类不存在无参数构造函数，则会编译报错。所以在类中添加有参数构造方法时，同时也应添上无参数构造方法。

（6）子类创建子类对象，CheckingAccount ca1＝new CheckingAccount()；子类对象调用从父类继承的方法，访问父类成员变量。

（7）若子类重写父类成员变量，子类对象访问该方法时只能访问重写的方法。

8.2.2　实例 2：几何图形类

【实例要求】

某画图软件可以画圆形、圆柱和球形，并通过它们的各自属性计算面积、周长或体积。画圆形、圆柱和球形都有确定的一个点来标识图形的位置。根据上述描述抽象出圆形类、圆柱类、球形类，然后通过这三个类抽象出点类，并编写测试类测试这些类。

【实例源代码】

新建源文件 Point.java，源代码如下：

```java
public class Point {
  /* 功能描述:点类 */
  private double x;
  private double y;
  public double getX() {
    return x;
  }
  public void setX(double x) {
    this.x = x;
  }
  public double getY() {
    return y;
  }
  public void setY(double y) {
    this.y = y;
  }
  public Point(){}
  public Point(double x,double y){
    this.x=x;
    this.y=y;
  }
  public void getInfo(){
    System.out.println("位于点,x="+x+",y="+y+"处。");
  }
}
```

新建源文件 Circle.java,源代码如下:

```java
public class Circle extends Point{
  /* 功能描述:圆类 */
  private double r;
  public double getR() {
      return r;
  }
  public void setR(double r) {
      this.r = r;
  }
  public Circle(){}
  public Circle(double x,double y,double r)
  {
      super(x,y);
      this.r=r;
  }
  public double perimenter(){
      return Math.PI * 2 * r;
  }
  public double area(){
      return Math.PI * r * r;
      //return Math.PI * Math.pow(r, 2);
```

```java
    }
    public void getInfo(){
        System.out.println("圆形圆心位于点 x="+super.getX()+",y="+super.getY());
        System.out.println("半径:"+this.r+"\n周长:"+this.perimenter()+"面积:"+this.area());
    }
}
```

新建源文件 Cylinder.java，源代码如下：

```java
package experiment8;
public class Cylinder extends Circle {
    /* 功能描述:圆柱类 */
    private double h;
    public Cylinder(){}
    public Cylinder(double x,double y,double h,double r){
        super(x,y,r);
        this.h=h;
    }
    public double area(){
        return 2 * Math.PI * Math.pow(super.getR(), 2)+super.perimenter() * h;
    }
    public double volume(){
        return super.area() * h;
    }
    public void getInfo(){
        System.out.println("圆柱底面圆心位于点 x="+super.getX()+",y="+super.getY());
        System.out.println("半径:"+super.getR()+"\n 高:"+h+"\n 体积:"+this.volume()+"\n 面积:"+this.area());
    }
}
```

新建源文件 Sphere.java，源代码如下：

```java
package experiment8;
public class Sphere extends Circle {
    /* 功能描述:球体类 */
    private double z;
    public Sphere(){}
    public Sphere(double x,double y,double z,double r){
        super(x,y,r);
        this.z=z;
    }
    public double area(){
        return Math.PI * 4 * Math.pow(super.getR(), 2);
    }
```

```
    public double volume(){
        return Math.PI * 4/3 * Math.pow(super.getR(), 3);
    }
    public void getInfo(){
        System.out.println("球形圆心位于点 x="+super.getX()+",y="+super.getY()+" z="+z);
        System.out.println("半径:"+super.getR()+"\n体积:"+this.volume()+"面积:"+this.area());
    }
}
```

新建源文件 CircleTest.java,源代码如下:

```
package experiment8;
public class CircleTest {
    /* 功能描述:测试类 */
    public static void main(String[] args) {
        System.out.println("------点------------------");
        Point p=new Point(1,2);
        p.getInfo();
        System.out.println("------圆------------------");
        Circle c=new Circle(1,2,10);
        c.getInfo();
        System.out.println("------球------------------");
        Sphere s=new Sphere(1,2,3,10.0);
        s.getInfo();
        System.out.println("------圆柱----------------");
        Cylinder cy =new Cylinder(1,3,8,10.0);
        cy.getInfo();
    }
}
```

程序运行结果如图 8-4 所示。

```
------点------------------
位于点, x=1.0, y=2.0处。
------圆------------------
圆形圆心位于点x=1.0,y=2.0
半径:10.0
周长:62.83185307179586面积:314.1592653589793
------球------------------
球形圆心位于点x=1.0,y=2.0z=3.0
半径:10.0
体积:4188.790204786391面积:1256.6370614359173
------圆柱----------------
圆柱底面圆心位于点x=1.0,y=3.0
半径:10.0
高:8.0
体积:2513.2741228718346
面积:1130.9733552923256
```

图 8-4 程序运行结果

【实例解析】

（1）根据题目要求，核心类是圆类、圆柱类和球形类，分析这三个类的共同属性是圆类。根据题目要求这三个类都存在点这个类，所以抽象出点类。本实例存在点类、圆类、圆柱类、球类。其中，圆类继承圆类、圆柱类和球类继承圆类。

（2）点类确定图像在画板的相对位置，所以存在 x、y 两个属性。该类存在一个方法获得点的信息，包括点的位置 public void getInfo()。

（3）圆类继承点类，并添加半径的属性，添加计算周长、面积的方法。重写获得信息的方法是 public void getInfo()。

（4）圆类方法的圆周率中使用数学类 Math 中的静态变量 PI，也可以在类中添加 public static final double PI＝3.14。

（5）圆柱类和球类继承圆类，并添加相应的属性，圆柱类和球类添加计算体积的方法，重写 public double area()、public void getInfo()。

（6）圆柱类和球类不用计算周长，所以圆类中计算周长的方法在圆柱类和球类不被使用。子类不可删除父类的属性和方法，可以覆盖和重写属性和方法。

（7）Java 语言语法规定不允许多重继承，允许多层继承。一个类不允许存在多个父类，一个父类可以存在多个子类。Super 关键字只能访问到当前类的父类。

8.3 上机实验

【实验目的】

- 理解继承的概念，掌握继承的语法。
- 了解重载与重写的区别。
- 掌握 super 关键字。
- 了解多态的概念。
- 掌握方法调配原则。

【实验要求】

某个公司要开发工资管理系统。该公司有多种类型的员工，不同类型的员工按照如下不同的方法计算月工资。

（1）总经理级，工资按照年薪计算，每月工资是年薪/12。

（2）经理级，基本工资＋补助＋工龄工资＋职务级别工资。

（3）普通员工级，基本工资＋补助＋工龄工资＋学历工资。

（4）销售人员级，基本工资＋补助＋工龄工资＋提成。

根据上述描述使用类的继承及相关机制设计类并编写一个 Java 程序演示它们。

实验运行效果如图 8-5 所示。

```
Problems * Javadoc * Declaration * Console *
<terminated> EmployeeTest [Java Application] D:\java\jdk1.7.0_72\bin\javaw.exe (2019年2月15日 下午4:46:23)
姓名：张三   身份证号：102311199002013214   工龄：3
底薪：3000.0   补助：1000.0   学历工资：4000   总工资：8300.0
姓名：李四   身份证号：102311199002013215   工龄：1
底薪：3000.0   补助：2000.0   提成：1500.0   总工资：6600.0
姓名：王五   身份证号：102311199002013216   工龄：5
底薪：3000.0   补助：4000.0   职务级别工资：5000.0   总工资：12500.0
姓名：赵六   身份证号：10231119900201321X   工龄：5
年薪：500000.0   本月发放：41666.666666666664
```

图 8-5 实验运行效果

【实验指导】

根据题目要求描述应先设计一个员工类（Employee），描述所有雇员的基本信息：姓名、身份证号、工龄、底薪、补助工资；该类包含 getter() 和 setter() 方法、构造方法、计算工资的方法、返回员工信息的方法。总经理、经理、普通员工、销售人员的类继承该员工类，并根据自身的特点添加相应属性，重写工资计算方法和返回信息方法。

【程序模板】

新建源文件 Employee.java，源代码如下：

```java
package salarymanagement;
public class Employee {
    [代码 1]
    private double salarybase;              //底薪
    private double salarysub;               //补助
    public String getName() {
        return name;
    }
    public void setName(String name) {
        this.name = name;
    }
    public String getID_card() {
        return id_card;
    }
    public void setID_card(String id_card) {
        this.id_card = id_card;
    }
    public int getWork_age() {
        return work_age;
    }
    public void setWork_age(int work_age) {
        this.work_age = work_age;
    }
    public double getSalarybase() {
        return salarybase;
```

```
    }
    public void setSalarybase(double salarybase) {
      this.salarybase = salarybase;
    }
    public double getSalarysub() {
      return salarysub;
    }
    public void setSalarysub(double salarysub) {
      this.salarysub = salarysub;
    }
    public Employee(String name, String id_card, int work_age, double salarybase,
double salarysub) {
      this.name = name;
      this.id_card = id_card;
      this.work_age = work_age;
      this.salarybase = salarybase;
      this.salarysub = salarysub;
    }
    public Employee(){}
    public double computersalary(){
      return [代码2];
    }
    public String toString(){
      return "姓名:"+this.name+" 身份证号:"+this.id_card+" 工龄:"+this.work_age+" ";
    }
}
```

新建文件GenerManager.Java，源程序如下：

```
package salarymanagement;
public class GenerManager[代码3]{
    private double yearsalary;              //年薪
    public double computersalary(){
        return [代码4];
    }
    public GenerManager(String name, String id_card, int work_age, double
salarybase, double salarysub, double yearsalary) {
        [代码5]
        this.yearsalary = yearsalary;
    }
    public String toString(){
        return super.toString()+"\n 年薪:"+this.yearsalary+" 本月发放:"+this.computersalary();
    }
}
```

新建源文件Manager.java，源程序如下：

```java
package salarymanagement;
public class Manager extends Employee{
    private double salarduties;              //职务级别工资
    public double getDuties() {
        return salarduties;
    }
    public void setDuties(double salarduties) {
        this.salarduties =salarduties;
    }
    public Manager(String name, String id_card, int work_age, double salarybase,
double salarysub,double salarduties) {
        super(name, id_card, work_age, salarybase, salarysub);
        this.salarduties=salarduties;
    }
    public double computersalary(){
        return super.computersalary()+salarduties;
    }
    public String toString(){
        [代码 6]+"\n底薪:"+this.getSalarybase()+" 补助:"+this.getSalarysub()+" 职务级
别工资:"+salarduties+" 总工资:"+this.computersalary();
    }
}
```

新建源文件Salesman.java,源代码如下:

```java
package salarymanagement;
public class Salesman extends Employee {
    private double commision;
    private int qualtities;
    public double getCommision() {
        return commision;
    }
    public void setCommision(double commision) {
        this.commision =commision;
    }
    public int getQualtities() {
        return qualtities;
    }
    public void setQualtities(int qualtities) {
        this.qualtities =qualtities;
    }
    public Salesman(){}
    public Salesman(double commision, int qualtities) {
        super();
        this.commision =commision;
        this.qualtities =qualtities;
    }
    public Salesman(String name, String id_card, int work_age, double salarybase,
double salarysub,double commision, int qualtities) {
```

```java
            super(name,id_card,work_age,salarybase,salarysub);
            this.commision =commision;
            this.qualtities =qualtities;
        }
        public double computersalary(){
            [代码 7]
        }
        public String toString(){
            return super.toString()+"\n底薪:"+this.getSalarybase()+"补助:"+this.getSalarysub()+"提成:"+this.commision * this.qualtities+"总工资:"+this.computersalary();
        }
}
```

新建源文件 Work.java,源代码如下:

```java
package salarymanagement;
public class Work extends Employee {
    private int education;                    //学历(0~4)
    public int getEducation() {
        return education;
    }
    public void setEducation(int education) {
        this.education =education;
    }
    public double computersalary(){
        return super.computersalary()+this.education * 2000;
    }
    public Work() {
        super();
    }
    public Work(String name, String id_card, int work_age, double salarybase, double salarysub,int education) {
        [代码 8]
        this.education=education;
    }
    public String toString(){
        return super.toString()+"\n底薪:"+this.getSalarybase()+"补助:"+this.getSalarysub()+"学历工资:"+ this.getEducation() * 2000 +"总工资:"+ this.computersalary();
    }
}
```

新建源文件 EmployeeTest.java,源代码如下:

```java
package salarymanagement;
public class EmployeeTest {
    public static void main(String[] args) {
        Employee [] em=new Employee[4];
```

```
            em[0]=new Work("张三","102311199002013214",3,3000,1000,2);
            em[1]=new Salesman("李四","102311199002013215",1,3000,2000,500,3);
            em[2]=new Manager("王五","102311199002013216",5,3000,4000,5000);
            em[3]=new GenerManager("赵六","10231119900201321X",5,0,0,500000);
            for(int i=0;i<em.length;i++){
                System.out.println([代码 9]);
            }
        }
    }
```

思考：尝试把 Employee 类改写成抽象类，把计算工资方法设计成抽象方法。

8.4 拓展练习

【基础知识练习】

一、选择题

1. 某一个子类要继承一个父类，要使用关键字(　　)。
   A. import　　　　B. extends　　　　C. implements　　　　D. java

2. 类的 3 个重要特征是类的封装、多态和(　　)。
   A. 实现　　　　B. 继承　　　　C. 重写　　　　D. 重载

3. 以下关于继承的叙述正确的是(　　)。
   A. 在 Java 中类只允许单一继承
   B. 在 Java 中一个类只能实现一个接口
   C. 在 Java 中一个类不能同时继承一个类和实现一个接口
   D. 在 Java 中接口只允许单一继承

4. 父类 A 被子类 B 继承，A 类中能被 B 类中访问的是(　　)。

```
class A{
    private int i =20;
    protected int j =30;
    public int k =40;
    int h =50;
}
class B extends A{
    void f() {
    }
}
```

　　A. i　　　　　　B. j　　　　　　C. k　　　　　　D. h

二、填空题

1. 在 Java 语言中，定义子类时，使用关键字_____来给出父类名。

2. 被_____关键字修饰的类不能被继承，被_____关键字修饰的成员方法不能被重写。

3. 在 Java 语言中,通常使用方法的重载和覆盖实现类的_____。

4. 若子类声明了一个与父类的成员方法同名的成员方法,则子类不能继承父类中的_____,_____此时称子类的_____覆盖了父类的成员方法。

5. 阅读下列程序,程序运行的结果是_____。

```java
public class Z extends X {
    Y y = new Y();
    Z() {
        System.out.print("Z");
    }
    public static void main(String[] args) {
        new Z();
    }
}
class X {
    Y b = new Y();
    X() {
        System.out.print("X");
    }
}
class Y {
    Y() {
        System.out.print("Y");
    }
}
```

【上机练习】

编写一个 Java 应用程序,设计一个交通工具类 Vehicle,在该类中包含属性有车速、交通工具种类、车辆颜色、是否低碳出行、最多承载人数,核心方法是计算乘坐该交通工具花费多长时间。自行车、汽车、公交车三个类继承交通工具类。自行车出行最多乘坐 2 个人,是低碳出行方式。公交车出行是绿色出行方式,计算出行时间需要添加平均等车时间、票价 1～5 元不等。汽车出行是非低碳出行方式,每辆汽车都有自己的品牌,且都有自己的每公里的油耗。最后,写一个测试类测试这些类的功能。

程序运行结果如图 8-6 所示。

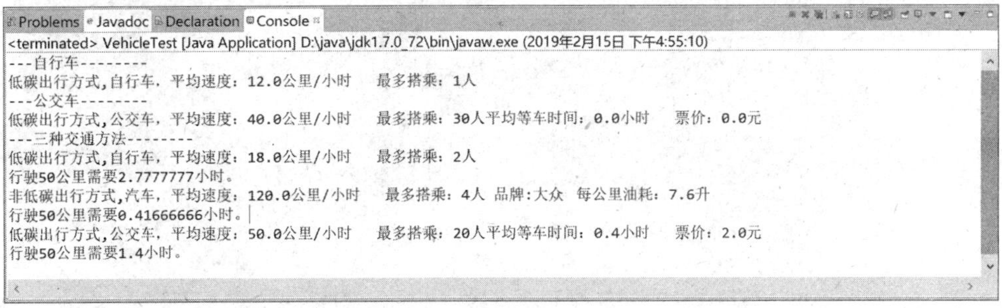

图 8-6 程序运行结果

提示：

（1）新建源文件 Vehicle.java，源代码如下：

```java
package vehicle;
public class Vehicle {
}
```

（2）新建源文件 Bicycle.java，源代码如下：

```java
package vehicle;
public class Bicycle extends Vehicle {
}
```

（3）新建源文件 Bus.java，源代码如下：

```java
package vehicle;
public class Bus extends Vehicle {
}
```

（4）新建源文件 Car.java，源代码如下：

```java
public class Car extends Vehicle {
}
```

（5）新建源文件 VehicleTest.java，源代码如下：

```java
package vehicle;
public class VehicleTest {
  public static void main(String[] args) {
      //TODO Auto-generated method stub
      Vehicle ve=new Bicycle();
      ve.setColor("蓝色");
      ve.setKind("自行车");
      ve.setLowCarbon(true);
      ve.setSpeed(12);
      ve.setNumber(1);
      System.out.println("---自行车--------");
      System.out.println(ve.getInfo());
      ve=new Bus();
      ve.setColor("绿色");
      ve.setKind("公交车");
      ve.setLowCarbon(true);
      ve.setSpeed(40);
      ve.setNumber(30);
      System.out.println("---公交车--------");
      System.out.println(ve.getInfo());
      System.out.println("---三种交通方法--------");
```

```
        Vehicle[] v=new Vehicle[3];
        v[0]=new Bicycle(18,"黄色");
        v[1]=new Car(120,"黑色",4,"大众",(float)7.6);
        v[2]=new Bus(50,"绿色",20,(float)0.4,2);
        for(int i=0;i<v.length;i++){
            System.out.println(v[i].getInfo());
            System.out.println("行驶50公里需要"+v[i].time(50)+"小时。");
        }
    }
}
```

第9章 接口与包

9.1 知识提炼

9.1.1 接口

接口是一组抽象方法、常量的集合。接口是指声明一套功能,没有具体的实现。接口定义的关键字 interface 的语法格式如下:

```
[public] interface InterfaceName [extends superInterfaceList]{
    //常量字段
    //抽象方法
}
```

说明:

(1) 接口只包括常量定义和抽象方法。

(2) 接口可以多重继承也可以多层继承,一个接口继承多个父类接口时,父类接口使用逗号分隔。

(3) 接口有 public 和默认两种访问修饰符,系统默认接口中所有属性的修饰符都是 public static final 及静态常量,系统默认接口中所有方法的修饰符都是 public abstract。

在开发平台中创建接口的步骤是:File→New→interface,弹出的对话框如图 9-1 所示。

Source folder:源文件夹的名称。

Package:源文件下的包。

Name:接口的名称。

Modifiers:接口的访问控制修饰符。

Extended interfaces:继承接口。

9.1.2 接口的实现

接口实现的关键字是 implement。

```
Class ClassName implement InterfaceList{
    //属性
    //构造方法
    //普通方法
    //接口抽象方法的实现
}
```

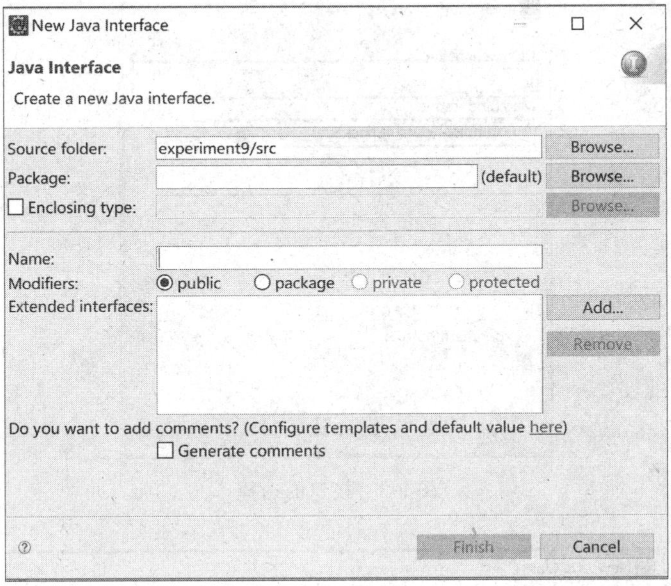

图 9-1 新建接口

接口实现类的定义格式如图 9-2～图 9-4 所示。

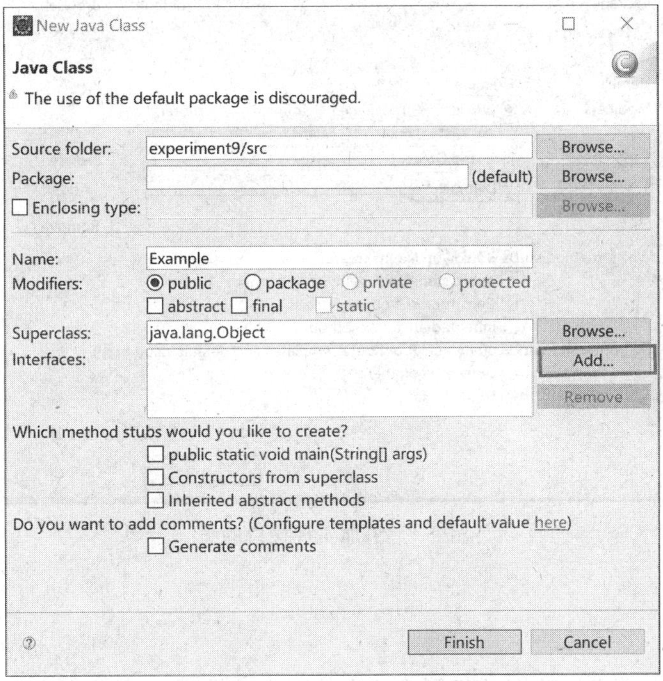

图 9-2 新建实现接口类

接口定义了一套行为规范，如果一个类要实现这个接口就要遵守接口中定义的规范，实际上就是要实现接口中定义的所有方法。接口可以多重继承，多层继承。一个类可以实现多个接口。类中实现接口的方法要加 public 修饰，因为接口中定义的抽象方法默认为 public。

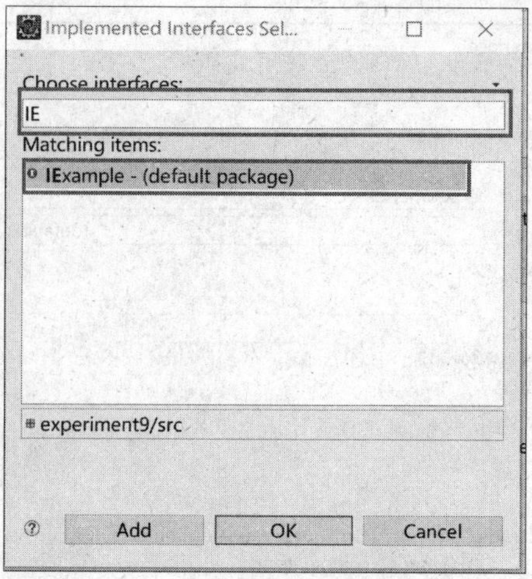

图 9-3 接口的选择

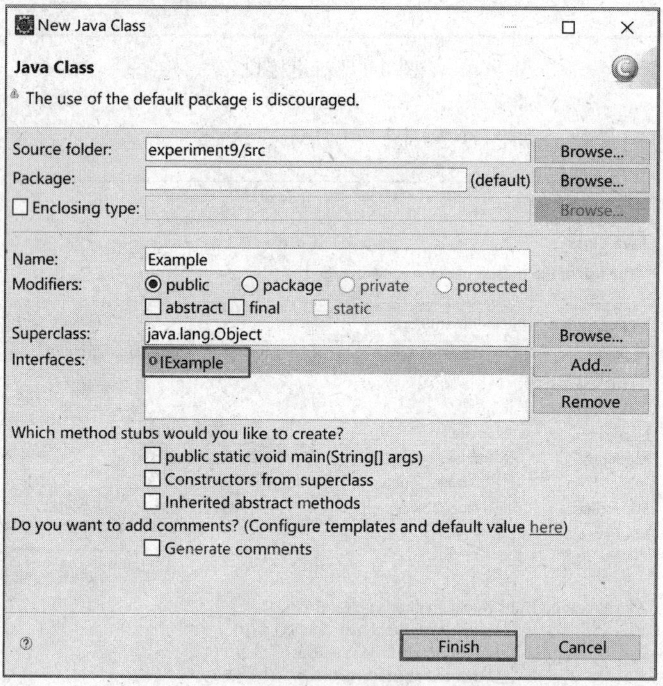

图 9-4 添加实现接口的类

9.1.3 包

包是 Java 语言中提供的一种区别于命名空间的机制。它是类的一种文件组织和管理方式,是一组功能相似或相关的类的接口的集合。使用关键字 package 定义一个包,其语法格式如下:

```
package packagename;
```

如同文件夹一样，包也采用树状目录的存储方式。同一个包中类的名字不能相同，不同包中类的名字可以相同。一个源程序只有一条 package 语句，且 package 语句必须放在源文件的首行，作为第一条非注释语句。包各层之间使用"."分隔，包的命名一般使用小写字母，避免使用与系统发生冲突的名字。

9.2 实例解析

9.2.1 实例1：求平均值

【实例要求】

在数学中有多种方法计算平均值，①在一组实数中全部数相加后求平均值；②去掉最大数、最小数，将剩余的数相加求平均值；③求加权平均数。要求使用 Java 语言的接口功能编写程序实现上述要求。具体要求如下：

（1）新建包 experiment9.average.Iaverage、experiment9.average.Caverage。

（2）在包 experiment9.average.Iaverage 中定义求平均值的接口。

（3）在包 experiment9.average.Caverage 中定义三个类分别实现求平均值的接口。这三个类分别实现一种求平均值的方法。

（4）在包 experiment9.average.Caverage 中定义测试类，测试上述三个类的功能和运行效果。测试类中给出一组数和一组权重，分别求得这组数的三种平均值。程序的运行结果如图 9-5 所示。

```
一组数：
10.0  8.0  9.0  7.0  8.0  9.0  10.0  6.0  9.0  10.0
平均数：8.6
去掉最高和最低分的平均数：8.75
权重：
0.2  0.1  0.05  0.05  0.1  0.1  0.2  0.05  0.05  0.1
加权平均数：9.05
```

图 9-5 求平均数

【实例源代码】

创建 experiment9.average.Iaverage 包，并在包中新建源文件 Average.java，源代码如下：

```
package experiment9.average.Iaverage;
public interface Average {
  /* 功能描述：声明平均数接口，声明获取平均数的方法 */
  public abstract double getAverage();
}
```

创建 experiment9.average.Caverage 包，并在包中新建源文件 CaverageA.java，源代码如下：

```java
package experiment9.average.Caverage;
import experiment9.average.Iaverage.Average;
public class CaverageA implements Average {
    /* 功能描述:求平均值类,把数组中所有数相加求平均值,实现求平均接口的方法 */
    private double [] number;
    public double [] getNumber() {
        return number;
    }
    public void setNumber(double [] number) {
        this.number =number;
    }
    public CaverageA(){}
    public CaverageA(double []number){
        this.setNumber(number);
    }
    public double getAverage() {
        double sum=0;
        for (double d : number) {
            sum+=d;
        }
        return sum/number.length;
    }
}
```

在包 experiment9.average.Caverage 中新建源文件 CaverageB.java,源代码如下:

```java
package experiment9.average.Caverage;
import experiment9.average.Iaverage.Average;
public class CaverageB implements Average {
    /* 功能描述:求平均值类,删除数组中的最大值和最小值后,所有数相加求平均值 */
//数组
    private double []number;
    public double [] getNumber() {
        return number;
    }
    public void setNumber(double[] number) {
        this.number =number;
    }

    public CaverageB(double[] number) {
        super();
        this.number =number;
    }
    @Override
    public double getAverage() {
        double max,min;
        double sum=0;
```

```
        max=min=number[0];
        for (double d : number) {
            if(max<d)
            {
                max=d;
            }
            if(min>d){
                min=d;
            }
            sum+=d;
        }
        return (sum-min-max)/(number.length-2);
    }
}
```

在包 experiment9.average.Caverage 中新建源文件 CaverageC.java,源代码如下:

```
package experiment9.average.Caverage;
import experiment9.average.Iaverage.Average;
public class CaverageC implements Average {
    /* 功能描述:求平均值类,在数组中求加权平均值 */
    private double[]number;
    private double[]weight;
    public double[] getNumber() {
        return number;
    }
    public void setNumber(double[] number) {
        this.number =number;
    }
    public double[] getWeight() {
        return weight;
    }
    public void setWeight(double[] weight) {
        this.weight =weight;
    }
    public CaverageC(double[] number, double[] weight) {
        super();
        this.number =number;
        this.weight =weight;
    }
    @Override
    public double getAverage() {
        double sum=0;
        for (int i =0; i <number.length; i++) {
            sum+=number[i] * weight[i];
        }
        return sum;
```

```
        }
    }
```

在包 experiment9.average.Caverage 中新建源文件 AverageTest.java,源代码如下:

```
package experiment9.average.Caverage;
import experiment9.average.Iaverage.Average;
public class AvergeTest {
    /* 功能描述:测试接口及实现接口的类 */
    public static void main(String[] args) {
        double []number=new double[]{10,8,9,7,8,9,10,6,9,10};
        double []weight=new double[]{0.2,0.1,0.05,0.05,0.1,0.1,0.2,0.05,0.05,0.1};
        System.out.println("一组数:");
        for (int i =0; i <number.length; i++) {
            System.out.print(number[i]+"  ");
        }
        System.out.println();
        Average ca=new CaverageA(number);
        System.out.println("平均数:"+ca.getAverage());
        ca=new CaverageB(number);
        System.out.println("去掉最高和最低分的平均数:"+ca.getAverage());
        System.out.println("权重:");
        for (double d : weight) {
            System.out.print(d+"  ");
        }
        System.out.println();
        ca=new CaverageC(number,weight);
        System.out.println("加权平均数:"+ca.getAverage());
    }
}
```

【实例解析】

(1) Java 项目开发时,为方便文件的管理,类似的文件或相同功能的文件放在一个包中。包的名称一般采用小写字母,包在编译时将转换成为目录层次。

(2) 若要使用某个包中的类、接口,需要使用关键字 import 将包导入源文件中。导入的内容放在 package 语句之后,类或接口的定义之前。

(3) 类使用 implements 关键字实现接口。

(4) 若接口里包含抽象方法,实现该接口的类必须实现接口声明的方法或者把实现的接口的类转化成抽象类。

(5) 接口不是类,不能创建对象,但是可以声明变量。本实例中使用 Average 接口声明变量 ca,通过 ca 调用实现该接口类的实例。接口变量只能调用接口中声明的方法。同样的调用方法 ca.getAverage()可以产生多种结果,实现多态。

(6) @Override 并不是关键字,但是可以把它当作关键字使用。当想要覆写(重写)某个方法时,可以选择添加这个注解,在不小心重载而并非覆写(重写)了该方法时,编译器就会生成一条错误信息。

9.2.2 实例2：计算面积与体积

【实例要求】

建立一个包 experiment9.cone.icomputer 并在包中分别定义两个接口，一个接口用于计算面积；另外一个接口用于计算体积。在另外一个包 experiment9.cone.test 中设计圆锥类和球类实现这两个接口，并在该包中编写相应的测试类测试实现接口的功能。

程序运行结果如图 9-6 所示。

```
圆锥的面积：471.23889803846896
球的面积：452.3893421169302
圆锥的体积：1047.1975511965977
球的体积：678.5840131753953
```

图 9-6　计算面积与体积接口测试结果

【实例源代码】

新建包 package experiment9.cone.icomputer，在包内新建源文件 Iarea.java，源代码如下：

```java
package experiment9.cone.icomputer;
public interface Iarea {
    /* 功能描述:计算面积接口,包含获得面积的抽象方法 */
    public double getArea();
}
```

在包 package experiment9.cone.icomputer 内新建源文件 Ivolume.java，源代码如下：

```java
package experiment9.cone.icomputer;
public interface Ivolume {
    /* 功能描述:计算体积接口,包含获得体积的抽象方法 */
    public double getVolume();
}
```

新建包 experiment9.cone.test，在包内新建源文件 Cone.java，源代码如下：

```java
package experiment9.cone.test;
import experiment9.cone.icomputer.Iarea;
import experiment9.cone.icomputer.Ivolume;
public class Cone implements Iarea, Ivolume {
    /* 功能描述:圆锥类,实现两个接口 */
    private double r;
    private double l;
    public double getR() {
```

```java
        return r;
    }

    public void setR(double r) {
        this.r = r;
    }
    public double getL() {
        return l;
    }

    public void setL(double l) {
        this.l = l;
    }
    public Cone() {}
    public Cone(double r,double l) {
        super();
        this.r = r;
        this.l = l;
    }

    @Override
    public double getVolume() {
        return Math.PI * r * r * r * 1/3;
    }
    @Override
    public double getArea() {
        return Math.PI * r * (r+l);
    }
}
```

在包 experiment9.cone.test 内新建源文件 Globe.java,源代码如下:

```java
package experiment9.cone.test;
import experiment9.cone.icomputer.Iarea;
import experiment9.cone.icomputer.Ivolume;
public class Globe implements Iarea, Ivolume {
    /* 功能描述:球体类实现两种接口 */
    private double r;
    public Globe(double r) {
        super();
        this.r = r;
    }
    public double getR() {
        return r;
    }
    public void setR(double r) {
```

```
            this.r = r;
        }
        @Override
        public double getVolume() {
            return 4/3 * Math.PI * r * r * r;
        }
        @Override
        public double getArea() {
            return 4 * Math.PI * r * r;
        }
}
```

在包 experiment9.cone.test 内新建源文件 Test.java，源代码如下：

```
package experiment9.cone.test;
import experiment9.cone.icomputer.*;
public class Test {
    /* 功能描述：测试接口及实现接口的类，传递接口参数 */
    public static void main(String[] args) {
        Cone co=new Cone(10,5);
        Globe cl=new Globe(6);
        System.out.print("圆锥的面积:");
        print((Iarea)co);
        System.out.print("球的面积:");
        print((Iarea)cl);

        System.out.print("圆锥的体积:");
        print((Ivolume)co);
        System.out.print("球的体积:");
        print((Iarea)co);
    }
    public static void print(Iarea ia){
        System.out.println(ia.getArea());
    }
    public static void print(Ivolume iv){
        System.out.println(iv.getVolume());
    }
}
```

【实例解析】

（1）一个类只能有一个直接基类，但是可以实现多个接口，从而弥补了 Java 语言不能多继承的问题。若一个类实现多个接口，就要实现多个接口中的抽象方法。

（2）接口类型可以声明变量，接口类型可以作为参数。实例中 public static void print (Iarea ia)，接口类型 Iarea 声明变量 ia，使用变量 ia 调用方法 getArea()。

（3）实例中两个方法是 public static void print(Iarea ia)、public static void print

(Ivolume iv)方法的重载。在两个方法中都使用了接口类型变量参数。Cone、Globe 两个类都实现了 Iarea 和 Ivolume 接口。这两个类的对象可以被接口 Iarea、Ivolume 声明的变量引用。

（4）若类实现某个接口，当该接口作为方法的参数时，调用该方法可使这个类的对象自动转化为该接口的类型。本实例因为方法的重载，所以需要进行强制转化。

9.3 上机实验

【实验目的】

- 了解接口的含义。
- 掌握接口声明、接口实现。
- 掌握包的概念、新建包、导入包。
- 掌握接口实现的多态性。

【实验要求】

某超市根据会员不同的积分给予不同的优惠策略。如若会员积分不足 3000 分不进行优惠活动；如果会员积分满 3000 分不足 6000 分且本次消费 500 元，则进行满 500 元减去 100 元的优惠；如果积分满 5000 分，则本次消费按照 8 折计算。优惠方式不能叠加。为了超市方便变更优惠策略，设计优惠策略接口。设计 3 种优惠方法类实现优惠策略接口。设计消费者类根据消费者的积分不同自动选择优惠方式。

测试用例见表 9-1。

表 9-1 测试用例

会员账号	姓名	密码	积分	本次消费金额
18288252708	张三	342123	4140	500
18288252712	李四	312123	2540	500
12134252712	赵五	213425	6540	400
12134252134	赵三	213425	6000	700

实验运行效果如图 9-7 所示。

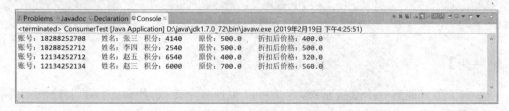

图 9-7 实验运行效果

【实验指导】

（1）首先创建两个包，一个包存放接口；另一个包存放类及测试类。

（2）本案例主要考查接口的设计及实现应用。首先设计优惠策略接口，优惠策略接口中声明一个获得优惠后价格的抽象方法，该方法的返回值是优惠后的价格，参数是消费金额。

（3）定义三个策略类实现优惠策略接口。三个类分别编写实现三种优惠策略是无优惠、满 500 元减 100 元、8 折优惠。

（4）定义消费者类。消费者类包含属性有会员账号、姓名、密码、积分、本次消费、优惠策略。核心方法是选择优惠策略计算优惠后的金额。

（5）编写相应的测试类测试上述程序。

【程序模板】

新建包 package experiment9.discount.Istrategy，在包内新建源文件 Istrategy.java，源代码如下：

```
package experiment9.discount.Istrategy;
public interface Istrategy {
  /* 功能描述：
   * 作者：
   * 日期：
   */
    [代码 1];
}
```

新建包 experiment9.discount.Cstrategy，在包内新建源文件 DiscountA.java，源代码如下：

```
package experiment9.discount.Cstrategy;
import experiment9.discount.Istrategy.*;
public class DiscountA[代码 2]{
  /* 功能描述：
   * 作者：
   * 日期：
   */
  @Override
  public double discountStrategy(double p) {
      [代码 3];
  }
}
```

在包 experiment9.discount.Cstrategy 中新建源文件 DiscountB.java，源代码如下：

```
package experiment9.discount.Cstrategy;
import experiment9.discount.Istrategy.Istrategy;
```

```
public class DiscountB implements Istrategy {
    /* 功能描述：
     * 作者：
     * 日期：
     */
    @Override
    public double discountStrategy(double p) {
        if([代码 4]){
            return p-100;
        }
        else{
            return p;
        }
    }
}
```

在包 experiment9.discount.Cstrategy 中新建源文件 DiscountC.java，源代码如下：

```
package experiment9.discount.Cstrategy;
import experiment9.discount.Istrategy.Istrategy;
public class DiscountC implements Istrategy {
    /* 功能描述：
     * 作者：
     * 日期：
     */
    @Override
    public double discountStrategy(double p) {
        [代码 5]
    }
}
```

在包 experiment9.discount.Cstrategy 中新建源文件 Consumer.java，源代码如下：

```
package experiment9.discount.Cstrategy;
import experiment9.discount.Istrategy.*;
public class Consumer {
    private String ID;
    private String name;
    private String password;
    private int integral;
    private double price;
    private Istrategy s;
    public String getID() {
        return ID;
    }
    public void setID(String iD) {
        ID = iD;
    }
    public String getName() {
```

```java
        return name;
    }
    public void setName(String name) {
        this.name =name;
    }
    public String getPassword() {
        return password;
    }
    public void setPassword(String password) {
        this.password =password;
    }
    public int getIntegral() {
        return integral;
    }
    public void setIntegral(int integral) {
        this.integral =integral;
    }
    public double getPrice() {
        return price;
    }
    public void setPrice(double price) {
        this.price =price;
    }
    public Istrategy getS() {
        return s;
    }
    public void setS() {
        int type=this.integral/1000;
        switch(type){
            case 0:
            case 1:
            case 2: [代码 6];break;
            case 3:
            case 4:
            case 5: [代码 7];break;
            default: [代码 8];break;
        }
    }
    public double GetResult(){
        return s.discountStrategy(this.price);
    }
    public Consumer(){}
    public Consumer(String iD, String name, String password, int integral, double price){
        ID =iD;
        this.name =name;
        this.password =password;
        this.integral =integral;
        this.price =price;
```

```
        this.setS();
    }
    public String toString(){
        return "账号:"+this.ID+"      姓名:"+this.name+"      积分:"+this.integral+"
原价:"+this.price+"     折扣后价格:"+this.GetResult();
    }
}
```

在包 experiment9.discount.Cstrategy 中新建源文件 ConsumerTest.java，源代码如下：

```
package experiment9.discount.Cstrategy;
public class ConsumerTest {
    public static void main(String[] args) {
        Consumer c1=new Consumer("18288252708","张三","342123",4140,500);
        System.out.println(c1.toString());
        Consumer c2=new Consumer("18288252712","李四","312123",2540,500);
        System.out.println(c2.toString());
        Consumer c3=new Consumer("12134252712","赵五","213425",6540,400);
        System.out.println(c3.toString());
        Consumer c4=new Consumer("12134252134","赵三","213425",6000,700);
        System.out.println(c4.toString());
    }
}
```

思考：若公司更改优惠策略，积分满 3000 分且不足 6000 分，会员采用 9 折优惠方式，且积分满 2000 分不到 3000 分的会员可参加满 500 元减 50 元的优惠策略。根据优惠策略的变化修改以上程序。

9.4 拓展练习

【基础知识练习】

一、选择题

1. 定义接口的关键字是（   ）。
   A. extends   B. class   C. interface   D. public

2. 下面关于 Java 的说法不正确的是（   ）。
   A. abstract 和 final 可以同时修饰一个类
   B. 抽象类不仅可以做父类，也可以做子类
   C. 抽象方法不一定声明在抽象类中，也可以在接口中
   D. 声明为 final 的方法不能在子类中覆写

3. 以下对接口描述不正确的有（   ）。
   A. 接口没有提供构造方法
   B. 接口中的方法默认使用 public、abstract 修饰
   C. 接口中的属性默认使用 public、abstract、final 修饰

D. 接口不允许多继承

4. 下列关于包、类和源文件的描述中，不正确的一项是（　　）。

　　A. 一个包可以包含多个类

　　B. 一个源文件中，只能有一个 public class

　　C. 属于同一个包的类在默认情况不可以互相访问，必须使用 import 导入

　　D. 系统不会为源文件创建默认的包

5. （　　）权限是同一包可以访问，不同包的子类可以访问，不同包的非子类不可以访问。

　　A. private　　　　　B. default　　　　　C. protected　　　　　D. public

二、填空题

1. 接口中的方法必须是_____，不能有方法体。

2. Java 不直接支持多继承，但可以通过_____实现多继承。

3. 下列程序运行的结果是_____。

```java
public interface InterfaceA {
  void print();
}
interface InterfaceB {
  void print();
}
public class Test_A_interface{
  public static void main(String[] args) {
      InterfaceA ia=new ADemo();
      ia.print();
  }
}
class ADemo implements InterfaceA,InterfaceB{
    @Override
    public void print() {
        //TODO Auto-generated method stub
        System.out.println("ADemo");
    }
}
```

4. 阅读下列程序，程序运行的结果是_____。

```java
public interface print {
    void printString();
}
public class Demo {
    public static void main(String[] args) {
        //TODO Auto-generated method stub
        print p=new A();
        p.printString();
        p=new B();
```

```
            p.printString();
        }
    }
    class A implements print{
        @Override
        public void printString() {
            //TODO Auto-generated method stub
            System.out.print("A");
        }
    }
    class B implements print{
        @Override
        public void printString() {
            //TODO Auto-generated method stub
            System.out.print("B");
        }
    }
```

【上机练习】

1. 计算器中最常用的四种运算方式为加、减、乘和除。尝试使用 Java 语言定义运算接口,定义加、减、乘、除四个类,使用加、减、乘、除四个类实现该运算接口并获得该类的运算结果。创建运算操作类,传入两个操作数和运算符号,根据运算符号的不同创建运算操作对象,使用运算对象调用方法求得这两个操作数在该运算作用下的结果。编写测试类程序测试该类。程序运行结果如图 9-8～图 9-11 所示。

```
输入运算符号:
+
输入数据:
90
100
计算结果: 90.0+100.0=190.0
```

图 9-8 加法运行结果

```
输入运算符号:
-
输入数据:
90
89
计算结果: 90.0-89.0=1.0
```

图 9-9 减法运行结果

```
输入运算符号:
*
输入数据:
8
7.4
计算结果: 8.0*7.4=59.2
```

图 9-10 乘法运行结果

```
输入运算符号:
/
输入数据:
100
0
计算结果: 100.0/0.0=1.7976931348623157E308
```

图 9-11 除法运行结果

2. 完成上题后,尝试添加求平方的功能,体会接口的优越性。

分析:

(1) 首先创建两个包,一个存放接口;另一个存放类。

(2) 编写运算接口 Ioperate,该接口中存放抽象方法 public abstract double getResult

（double number1,double number2）。

(3) 编写加、减、乘、除四个类实现运算接口。

(4) 编写运算类 CreateOperation，在该类中存在四个成员变量：Ioperate 接口变量、运算符号、两个操作数。

(5) 编写测试类测试上述功能。

文件读/写

10.1 知识提炼

10.1.1 File 类

文件(File)是最常见的数据源之一,借助 File 对象可以获取文件或相关目录的信息。File 类提供了与具体平台无关的方式操作文件。File 类常用的方法见表 10-1。

表 10-1 File 类的常用方法

方法	功能
boolean createNewFile()	创建文件
boolean delete()	删除文件或删除文件夹
boolean exists()	判断当前文件夹或文件是否存在
String getName()	获取文件或文件夹的名称
String getAbsolutePath()	获得当前文件或文件夹的绝对路径
boolean isDirectory()	判断是否是目录
boolean isFile()	判断是否是文件
long length()	文件的实际大小
String[] list()	返回文件夹所有的文件名称和文件夹名
File[] listFiles()	返回当前文件夹下所有文件对象
boolean mkdir()	创建当前文件的子目录
boolean equals(File f)	比较文件夹或目录是否相等

10.1.2 字节流

InputStream 类和 OutputStream 类是抽象类,是面向字节的输入/输出流的根。这两个类的主要方法见表 10-2 和表 10-3。

表 10-2 InputStream 类的常用方法

方法	功能
Int read()	读一个字节
Int read(byte b[])	读多个字节到字节数组
Int read(byte[] b, int off, int len)	读指定长度的数据到字节数组

续表

方　　法	功　　能
Void close()	关闭流
Long skip(long n)	输入当前指针的位置
Void mark()	在当前位置做一处标记

表 10-3　OutputStream 类的常用方法

方　　法	功　　能
Void write(int b)	将参数的低字节写入输入流
Void write(byte b[])	将字节数组全部写入输出流
Void write(byte b[],int offset,int len)	将字节数组从 b[offset]开始共 len 个字节写入输入流
Void close()	关闭流

（1）FileInputStream 类和 FileOutputStream 类分别直接继承于 InputStream 类和 OutputStream 类，继承或重写父类的方法用于面向字节流的文件读/写。

（2）DataInputStream 类和 DataOutputStream 类是实现各种数据类型的输入/输出。这两个类的对象必须和一个输入类或输出类联系起来，不能直接使用文件名或文件对象作为参数创建对象。

（3）BufferedInputStream 类和 BufferedOutputStream 类是缓冲流类，可避免每个字节的读/写对流进行。缓冲数据流的对象创建必须使用 FileInputStream 类或 FileOutputStream 类的对象作为参数。

10.1.3　字符流

Reader 类和 Writer 类是面向字符的输入/输出流类，是抽象类。与 InputStream 类和 OutputStream 类相同，只是方法的参数类型从 byte 改为 char。

（1）InputStreamReader 类和 OutputStreamWriter 类是处理字符流的基本类。

（2）BufferedReader 类和 BufferedWriter 类是缓冲字符流类，可提高字符的处理效率。

10.1.4　随机访问文件

RandomAccessFile 是功能丰富的文件访问类，既可以读文件也可以写文件，还支持随机访问的方式。该类的主要方法及功能介绍如表 10-4 所示。

表 10-4　RandomAccessFile 类的常用方法

方　　法	功　　能
RandomAccessFile(File file,String mode)	构造方法
RandomAccessFile(String name,String mode)	构造方法
Int read()	读一个字节
Int read(byte b[])	读多个字节到字节数组
Int read(byte[] b,int off,int len)	读指定长度的数据到字节数组
Void write(int b)	将参数的低字节写入输入流

续表

方　　法	功　　能
Void write(byte b[])	将字节数组全部写入输出流
Void write(byte b[],int offset,int len)	将字节数组从 b[offset]开始共 len 个字节写入输入流
Long getFilePointer()	返回文件指针的位置
Void seek(long pos)	将文件指针定位到一个绝对位置

10.1.5　对象序列化

对象输入/输出流类 ObjectInputStream 类和 ObjectOutputStream 类将 Java 文件读/写流扩充到对象的输入/输出。ObjectInputStream 类和 ObjectOutputStream 类提供 readObject()和 writeObject()方法实现了对象的反串行化和串行化。

若对象实现序列化,类必须实现 Serializable 接口或 Externalizable 接口。Serializable 接口是标记接口,易于实现且存储空间较大,但成本较高且速度较慢。Externalizable 接口存储空间小、成本低且速度快,但较难实现。

10.2　实例解析

10.2.1　实例1:完全数文件读/写

【实例要求】

完全数又称完美数或完备数,是一些特殊的自然数。它所有的真因子(即除了自身以外的约数)的和(即因子函数)恰好等于它本身。如果一个数恰好等于它的因子之和,则称该数为完全数。第一个完全数是 6,它的约数是 1、2、3、6,除去它本身 6 外,其余 3 个数相加,1+2+3=6。第二个完全数是 28,它的约数是 1、2、4、7、14、28,除去它本身 28 外,其余 5 个数相加,1+2+4+7+14=28。求出 1 到 10000 之间的完全数,使用字节流和数据流分别写入文件并输出到控制台,并从文件中读出完全数输出到控制台。程序运行如图 10-1 所示。

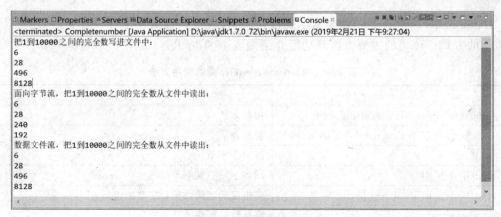

图 10-1　简单问答程序

【实例源代码】

新建源文件 Completenumber.java,源代码如下:

```java
import java.io.DataInputStream;
import java.io.DataOutputStream;
import java.io.FileInputStream;
import java.io.FileOutputStream;
import java.io.IOException;
public class Completenumber {
    /* 功能描述: 素数读/写 */
    public static void main(String[] args) throws IOException {
        FileOutputStream fos=new FileOutputStream("D:/completenumberbyte.txt");
        FileOutputStream fout=new FileOutputStream("D:/completenumberdata.txt");
        DataOutputStream dout=new DataOutputStream(fout);
        System.out.println("把 1 到 10000 之间的完全数写进文件中:");
        for(int i=1;i<=10000;i++) {
            int t =0;
            for(int j=1;j<=i/2;j++) {
                if(i%j==0) {
                    t+=j;
                }
            }
            if(t==i) {
                //使用字节文件流写入文件 D:/completenumberbyte.txt
                fos.write(i);
                //使用数据文件流写入文件 D:/completenumberdata.txt
                dout.writeInt(i);
                System.out.println(i);
            }
        }
        fos.close();
        dout.flush();
        dout.close();
        //从文件夹 D:/completenumberbyte.txt 中使用面向字节流类读文件
        System.out.println("面向字节流,把 1 到 10000 之间的完全数从文件中读出:");
        FileInputStream fis=new FileInputStream("D:/completenumberbyte.txt");
        int temp;
        while((temp=fis.read())!=-1){
            System.out.println(temp);
        }
        fis.close();
        //从文件夹 D:/completenumberdata.txt 中使用数据文件流读文件
        System.out.println("数据文件流,把 1 到 10000 之间的完全数从文件中读出:");
        FileInputStream fin=new FileInputStream("D:/completenumberdata.txt");
        DataInputStream din=new DataInputStream(fin);
        int filemax=din.available();
        for (int i=0; i<filemax; i+=4) {
```

```
            System.out.println(din.readInt());
        }
        din.close();
    }
}
```

【实例解析】

(1) 字节流读/写文件的步骤。

① 使用 File 类指向某一个文件或确定文件地址。

② 通过字节流或字符流的子类,指定输出的位置。

③ 进行读/写操作。

④ 关闭输入/输出。

I/O 操作属于资源操作,一定要记得关闭。

(2) 文件读/写属于资源操作时可能会抛出 IOException 异常。文件异常不属于 RuntimeException。当使用文件操作时必须做出异常处理或异常抛出。

10.2.2 实例 2:电话号码提取

【实例要求】

文本文档 AddressBook.txt 中存放联系人的姓名、电话号码、个人地址。先使用 Java 编写程序从 AddressBook.txt 中读出文件,截取电话号码存放在 NewAddressBook.txt 内。AddressBook.txt 文件内容如下:

小张 18888800001 广东广州。

刘老师 14567809999 广东潮汕。

张三 14488887891 广东肇庆。

王老师 18945679999 广东深圳。

赵二 14545677899 广东佛山。

赵老师 12378901234 广东珠海。

刘哥 17890900001 广东肇庆。

NewAddressBook.txt 文件内容如下:

18888800001

14567809999

14488887891

18945679999

14545677899

12378901234

17890900001

【实例源代码】

新建源文件 Address.java,源代码如下:

```java
import java.io.BufferedReader;
import java.io.BufferedWriter;
import java.io.File;
import java.io.FileInputStream;
import java.io.FileWriter;
import java.io.IOException;
import java.io.InputStreamReader;
public class Address {
    /* 功能描述：电话号码提取 */
    public static void main(String[] args) {
        String s=readFile("D:\\java\\example\\experiment10\\AddressBook.txt");
        writerFile(s,"D:\\java\\example\\experiment10\\NewAddressBook.txt");
    }
    public static String readFile(String path){
        StringBuilder content=new StringBuilder("");
        try{
            String encoding="UTF-8";
            File file=new File(path);
            if(file.isFile()&&file.exists()){
                InputStreamReader read=new InputStreamReader(new FileInputStream(file),encoding);
                BufferedReader bufferedReader=new BufferedReader(read);
                String temp=null;
                while((temp=bufferedReader.readLine())!=null){
                    String[]result=divName(temp);
                    content.append(result[1]+"\r\n");
                }
                read.close();
            }else{
                System.out.println("找不到文件!");
            }
        }catch(Exception e){
            System.out.println("文件操作有问题!");
            e.printStackTrace();
        }
        return content.toString();
    }
    private static String[] divName(String temp) {
        String[]result=new String[2];
        int index=0;
        for (int i =0; i <temp.length(); i++) {
            if(temp.charAt(i)>='0'&&temp.charAt(i)<='9'){
                index=i;
                break;
            }
```

```
        }
        result[0]=temp.substring(0,index);
        result[1]=temp.substring(index,index+11);
        return result;
    }
    private static void writerFile(String content,String path){
        BufferedWriter bw=null;
        try{
            File file=new File(path);
            if(!file.exists()){
                file.createNewFile();
            }
            FileWriter fw=new FileWriter(file.getAbsoluteFile());
            bw=new BufferedWriter(fw);
            bw.write(content);
            bw.close();
        }catch(IOException e){
            e.printStackTrace();
        }
    }
}
```

【实例解析】

（1）为了提高字符流读/写的效率，引入了缓冲机制，进行字符批量的读/写，提高了单个字符读/写的效率。BufferedReader 用于加快读取字符的速度，BufferedWriter 用于加快写入的速度。

（2）readLine()：直到程序遇到了换行符或者是对应流的结束符，该方法才会认为读完了一行，才会结束其阻塞，让程序继续往下执行。

（3）append()：append()方法的作用是在一个 StringBuffer 对象后面追加字符串。

（4）substring()是一个重载方法。public String substring(int beginIndex)返回一个新字符串，它是此字符串的一个子字符串。该子字符串始于指定索引处的字符，一直到此字符串末尾。public String substring(int beginIndex, int endIndex)返回一个新字符串，它是此字符串的一个子字符串。该子字符串从指定的 beginIndex 处开始，到指定的 endIndex−1 处结束。

10.2.3 实例3：字数统计

【实例要求】

在项目中新建 txt 文件，存放《春江花月夜》这首古诗。使用 Java 语言编写程序识别春江花月夜.txt 文件中存在几个"江"字。程序运行结果如图 10-2 所示。

【实例源代码】

新建源文件 Statistics.java，源代码如下：

图 10-2　程序运行结果

```java
import java.io.BufferedReader;
import java.io.FileReader;
import java.io.IOException;
public class Statistics {
  public static int findWord(String s,char ch){
    int n=0;
    int counter=0;
    while(n!=-1){
        n=s.indexOf(ch, n+1);
        counter++;
    }
    return counter-1;
  }
  public static void main(String[] args){
    try{
        FileReader fr=new FileReader(args[0]);
        BufferedReader brin=new BufferedReader(fr);
        String s="",x;
        while((x=brin.readLine())!=null){
            s=s+x;
        }
        brin.close();
        fr.close();
        char c='江';
        System.out.println(args[0]+"中共含有"江"的个数是:");
        System.out.println(findWord(s,c));
    }catch(IOException e){
    };
  }
}
```

【实例解析】

（1）在项目中添加文件,如图 10-3～图 10-5 所示。

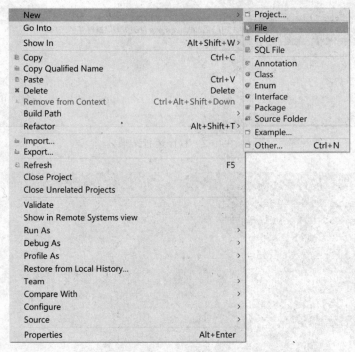

图 10-3 在项目中添加文件

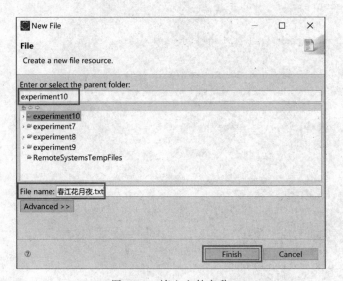

图 10-4 输入文件名称

图 10-5 刷新出现文件列表

（2）主函数有参数 FileReader fr＝new FileReader(args[0])；如何在控制台运行有参数的主方法？向参数添加数据，右击→Run As→Run Configurations，如图 10-6 和图 10-7 所示。

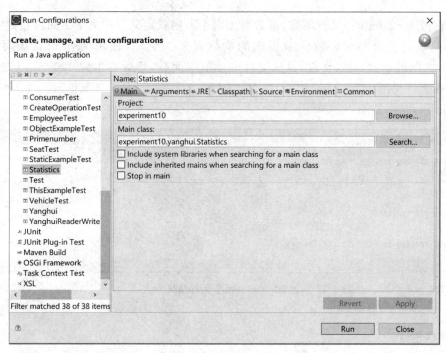

图 10-6　将参数输入对话框

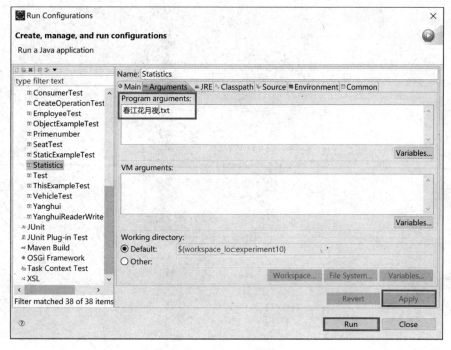

图 10-7　输入参数后确定

10.2.4 实例4：素数的随机读/写

【实例要求】

素数(prime number)又称质数，素数有无限个。质数定义为在大于1的自然数中，除了1和它本身以外没有其他因数。10以内的素数有2、3、5、7。使用Java语言编写程序求100~120之间的素数，写入文件，并从文件中读取中间数，然后再生成120~150之间的素数，追加在文件尾部。

程序运行结果如图10-8所示。

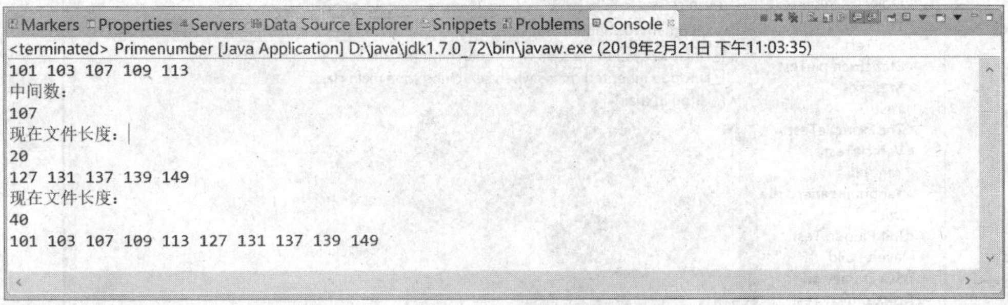

```
101 103 107 109 113
中间数：
107
现在文件长度：
20
127 131 137 139 149
现在文件长度：
40
101 103 107 109 113 127 131 137 139 149
```

图10-8 程序运行结果

【实例源代码】

新建源文件 Primenumber.java，源代码如下：

```java
import java.io.IOException;
import java.io.RandomAccessFile;
public class Primenumber {
  /* 功能描述：素数产生及随机读/写 */
  public static void main(String[] args) {
    try {
      RandomAccessFile raf=new RandomAccessFile("Primenumber.txt","rw");
      int count=0;
      for(int i=101;i<200;i+=2) {
        boolean flag=true;
        for(int j=2;j<=Math.sqrt(i);j++) {
          if(i%j==0) {
            flag=false;
            break;
          }
        }
        if(flag==true) {
          count++;
          raf.writeInt(i);
          System.out.print(i+"\t");
```

```
            }
        }
        System.out.println();
        raf.seek((count/2) * 4);
        System.out.println(raf.readInt());
        raf.close();
        RandomAccessFile rbf=new RandomAccessFile("Primenumber.txt","rw");
        rbf.seek(rbf.length());
        for(int i=200;i<300;i+=1) {
            boolean flag=true;
            for(int j=2;j<=Math.sqrt(i);j++) {
                if(i%j==0) {
                    flag=false;
                    break;
                }
            }
            if(flag==true) {
                count++;
                rbf.writeInt(i);
                System.out.print(i+"\t");
            }
        }
        System.out.println();
        for (int i =0; i <rbf.length(); i+=4) {
            rbf.seek(i);
            System.out.print(rbf.readInt()+"\t");
        }
        rbf.close();
    } catch (IOException e) {
        //TODO Auto-generated catch block
        e.printStackTrace();
    }
  }
}
```

【实例解析】

(1) Math.sqrt(i): 是 Math 的静态方法用来求开平方的结果。

(2) RandomAccessFile 类: RandomAccessFile 是 Java 输入/输出流体系中功能最丰富的文件内容访问类,既可以读取文件内容,也可以向文件输出数据。与普通的输入/输出流不同的是,RandomAccessFile 支持跳到文件任意位置读/写数据,RandomAccessFile 对象包含一个记录指针,用来标识当前读/写处的位置,当程序创建一个新的 RandomAccessFile 对象时,该对象的文件记录指针在文件头(也就是 0 处),当读/写 n 字节后,文件记录指针将会向后移动 n 字节。

(3) RandomAccessFile("Primenumber.txt","rw"): RandomAccessFile 类在创建对

象时,除了指定文件本身,还需要指定一个 mode 参数,该参数指定 RandomAccessFile 的访问模式,该参数有以下四个值。

- r:以只读方式打开指定文件。如果试图对该 RandomAccessFile 指定的文件执行写入操作则会抛出 IOException。
- rw:以读取、写入方式打开指定文件。如果该文件不存在,则尝试创建文件。
- rws:以读取、写入方式打开指定文件。相对于 rw 模式,还要求对文件的内容或元数据的每个更新都同步写入底层的存储设备。默认情形下(rw 模式下)使用 buffer,只有 cache 满或者使用 RandomAccessFile.close()关闭流的时候才真正写入文件中。
- rwd:与 rws 类似,只是仅将文件的内容同步更新到磁盘,而不修改文件的元数据。

(4) raf.seek((count/2) * 4)和 void seek(long pos)方法是将文件记录指针定位到 pos 位置定位指针,指针移动的基本单位是字节。

(5) rbf.length():获得文件的长度。

(6) rbf.writeInt(i):向文件中写入一个整型数,4 字节。rbf.readInt()用于从文件中读取一个整型数。readXxx()和 writeXxx()用于处理各种值类型。

(7) getFilePointer():获取当前文件的记录指针位置。read()和 write()分别为读取内容和写入内容。

10.3 上机实验

【实验目的】

- 理解流的概念。
- 掌握文件读/写的一般步骤。
- 掌握对象序列化概念。
- 编写并实现对象的读/写。

【实验要求】

使用 Java 语言编写学生成绩管理程序。假设本学期学生主修英语、高数和 Java 3 门课程,设计文件管理程序,录入指定学生的个人信息及成绩,将学生的成绩存放在文件中。计算每位学生的平均分并计算高数、英语、Java 的各科平均分。

测试用例见表 10-5。

表 10-5 测试用例

学 号	姓名	性别	高数	英语	Java
123456789	张三	男	70.0	75.0	80.0
123456788	赵六	女	80.0	90.0	90.0
123456787	李二	男	80.0	90.0	90.0

实验运行结果如图 10-9 所示。

```
第3个学生：
学号：
123456787
姓名：
李二
性别（1代表男生，2代表女生）：
1
高数成绩：
80
英语成绩：
90
Java成绩：
90
学生成绩信息已经读入文件！
从文件中读取学生成绩信息如下：
学号：123456789 姓名：张三 性别：男 高数：70.0 英语：75.0 Java:80.0
学号：123456788 姓名：赵六 性别：女 高数：80.0 英语：90.0 Java:90.0
学号：123456787 姓名：李二 性别：男 高数：80.0 英语：90.0 Java:90.0
高数平均成绩：76.66666666666667
英语平均成绩：85.0
Java平均成绩：86.66666666666667
```

图 10-9 实验运行结果

【实验指导】

按照题目要求程序应有以下关键步骤。

（1）建立学生类(学号、姓名、性别、各科成绩)，各个属性的 getter() 和 setter() 方法、构造方法以及求 3 门课程成绩平均分方法。

（2）录入指定学生个数和学生信息并写入文件中。

（3）从文件中读出学生信息放进数组，并计算平均成绩。

（4）计算并输出各科成绩的平均分。

（5）对象序列化。

【程序模板】

新建源文件 Student.java，源代码如下：

```java
import java.io.Serializable;
public class Student implements Serializable {
    private String ID;                    //学号
    private String name;                  //姓名
    private byte sex;                     //性别 1代表男,2代表女
    private double gaoshu;                //数学成绩
    private double yingyu;                //英语成绩
    private double java;                  //Java 成绩
    //getter 和 setter 方法
    public String getID() {
        return ID;
    }
    public void setID(String iD) {
        ID = iD;
    }
    public String getName() {
```

```java
        return name;
    }
    public void setName(String name) {
        this.name = name;
    }
    public char getSex() {
        if([代码 1]){
            return '男';
        }else if([代码 2]){
            return '女';
        }else{
            return ' ';
        }
    }
    public void setSex(byte sex) {
        this.sex = sex;
    }
    public double getGaoshu() {
        return gaoshu;
    }
    public void setGaoshu(double gaoshu) {
        this.gaoshu = gaoshu;
    }
    public double getYingyu() {
        return yingyu;
    }
    public void setYingyu(double yingyu) {
        this.yingyu = yingyu;
    }
    public double getJava() {
        return java;
    }
    public void setJava(double java) {
        this.java = java;
    }
    //构造方法
    public Student(){}
    public Student([代码 3]) {
        super();
        ID = iD;
        this.name = name;
        this.sex = sex;
        this.gaoshu = gaoshu;
        this.yingyu = yingyu;
        this.java = java;
    }
    public double average(){
        [代码 4];
    }
```

```java
    public String toString(){
        return "学号:"+ID+" 姓名:"+name+" 性别:"+this.getSex()+" 高数:"+gaoshu+" 英语:"+yingyu+" Java:"+java;
    }
}
```

新建源文件 StudentFile.java,源代码如下:

```java
import java.io.BufferedReader;
import java.io.FileInputStream;
import java.io.FileOutputStream;
import java.io.IOException;
import java.io.InputStreamReader;
import java.io.ObjectInputStream;
import java.io.ObjectOutputStream;
public class StudentFile {
    public static String input(String outs){
        String s="";
        try{
            BufferedReader br=new BufferedReader(new InputStreamReader(System.in));
            System.out.println(outs);
            s=br.readLine();
        }catch(IOException e){};
        return s;
    }
    public static void main(String[] args) {
        int number=Integer.parseInt(input("请输入学生个数:"));
        Student[] stu=new Student[number];
        for (int i =0; i <stu.length; i++) {
            //创建学生对象
            stu[i] =[代码 5];
        }
        for (int i =0; i <stu.length; i++) {
            System.out.println("第"+(i+1)+"个学生:");
            stu[i].setID(input("学号:"));
            stu[i].setName(input("姓名:"));
            stu[i].setSex(Byte.parseByte(input("性别(1代表男生,2代表女生):")));
            stu[i].setGaoshu(Double.parseDouble(input("高数成绩:")));
            stu[i].setYingyu(Double.parseDouble(input("英语成绩:")));
            stu[i].setJava(Double.parseDouble(input("Java 成绩:")));
        }
        //写入文件
        try{
            ObjectOutputStream out=new ObjectOutputStream(new FileOutputStream("学生成绩文件.dat"));
            [代码 6]
            for (int i =0; i <stu.length; i++) {
                out.writeObject(stu[i]);
```

```
            }
            out.close();
            System.out.println("学生成绩信息已经读入文件!");
        }catch(IOException e){
            e.printStackTrace();
        }
        //读出文件
        try{
            ObjectInputStream in=new ObjectInputStream(new FileInputStream("学生成绩文件.dat"));
            int n=((Integer)in.readObject()).intValue();
            System.out.println("从文件中读取学生成绩信息如下:");
            double gaoshu_sum=0,yingyu_sum=0,java_sum=0;
            Student[] st=new Student[n];
            for (int i =0; i <stu.length; i++) {
                st[i]=(Student)in.readObject();
                System.out.println(st[i].toString());
                gaoshu_sum+=st[i].getGaoshu();
                yingyu_sum+=st[i].getYingyu();
                java_sum+=st[i].getJava();
            }
            System.out.println("高数平均成绩:"+(gaoshu_sum/n));
            System.out.println("英语平均成绩:"+(yingyu_sum/n));
            System.out.println("Java 平均成绩:"+(java_sum/n));
            in.close();
        }catch(IOException e){
            e.printStackTrace();
        }catch (ClassNotFoundException e) {
            //TODO Auto-generated catch block
            e.printStackTrace();
        }
    }
}
```

10.4 拓展练习

【基础知识练习】

一、选择题

1. File 类提供了许多管理磁盘的方法。其中,建立目录的方法是(　　　)。
   A. delete()　　　　B. mkdirs()　　　　C. makedir()　　　　D. exists()
2. InputStream 类提供了(　　　)流功能。
   A. 输入　　　　B. 输出　　　　C. 打开　　　　D. 关闭
3. 下列程序从标准输入设备读入一个字符,然后再输出到显示器,"//x"处可填写(　　　)完成要求的功能。

```
import java.io.*;
    public class X8_1_4 {
        public static void main(String[] args) {
            char ch;
            try{
                //x
                System.out.println(ch);
            }
            catch(IOException e){
                e.printStackTrace();
            }
        }
    }
```

  A. ch = System.in.read();   B. ch = (char)System.in.read();
  C. ch = (char)System.in.readln();   D. ch = (int)System.in.read();

  4. 下面的说法不正确的是( )。
  A. InputStream 与 OutputStream 类通常用来处理字节流,也就是二进制文件
  B. Reader 与 Writer 类是用来处理字符流,也就是纯文本文件
  C. Java 中 I/O 流的处理通常分为输入和输出两个部分
  D. File 类是输入/输出流类的子类

  5. 能读出字节数据进行 Java 基本数据类型判断过滤的类是( )。
  A. BufferedInputStream   B. FileInputStream
  C. DataInputStream   D. FileReader

二、填空题

  1. Java 的输入/输出流包括 _____ 、_____ 、_____ 、_____ 以及多线程之间通信的 _____ 。

  2. Java 语言的 java.io 包中的 _____ 类是专门用来管理磁盘文件和目录的。调用 _____ 类的方法则可以完成对文件或目录的常用管理操作,如创建文件或目录、删除文件或目录、查看文件的有关信息等。

  3. 文件输入流是 _____ 、_____ ,文件输出流是 _____ 、_____ 。

  4. I/O 操作中字节流的操作类是 _____ 、_____ ,字符流的操作类是 _____ 和 _____ 。

  5. 序列化对象使用 _____ 、_____ 类,对象所在的类必须实现 _____ 接口,才可以自动序列化所有的内容。

【上机练习】

  杨辉三角是二项式系数在三角形中的一种几何排列,在中国南宋数学家杨辉1261年所著的《详解九章算法》一书中出现。在欧洲,帕斯卡(1623—1662年)在1654年发现这一规律,所以这个表又叫作帕斯卡三角形。杨辉三角的规则如下:

- 每行的端点与结束都是1。
- 每个数等于它上方两数之和。

- 每行数字左右对称,由 1 开始逐渐变大。
- 第 n 行的数字有 n 项。

使用 Java 语言编写程序求出前 10 行的杨辉三角,使用面向字节流、字符流两种方法将杨辉三角写进文件,并用对应的方法从文件中读出杨辉三角。程序运行结果如图 10-10 所示。

提示:所学过文件读/写的方法皆可使用。

核心代码提示:

```
int[][] a =new int[10][10];
for(int i=0; i<10; i++) {
    a[i][i] =1;
    a[i][0] =1;
}
for(int i=2; i<10; i++) {
    for(int j=1; j<i; j++) {
        a[i][j] =a[i-1][j-1] +a[i-1][j];
    }
}
```

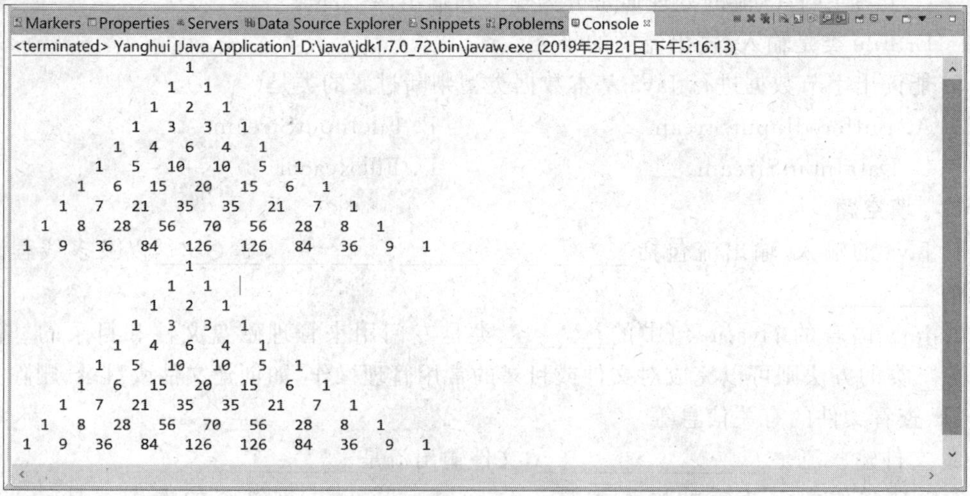

图 10-10 杨辉三角读/写程序运行结果

第 11 章

泛型与集合

11.1 知识提炼

11.1.1 泛型

泛型是 Java SE 1.5 版本中推出的一个新特性,它能够简化部分烦琐的编码方式,提高代码的安全性和重用率。具体就是:在定义类、接口、方法、方法的参数或成员变量时,指定它们的操作对象的类型为通用类型(即任意数据类型),而非具体的数据类型;在使用这些类、接口、方法、方法的参数或成员变量时,再将通用类型转换成指定的数据类型。

泛型定义语法格式如下:

```
Class MyExample<T>{}
```

在类名后面加一对尖括号,尖括号中用一个任意字符代表一个类型参数名称。

类型参数的名称可以任选,但是按照惯例,建议使用单个大写字母来表示,在 Java SE API 中常用的类型参数名如下。

E:表示集合中的元素类型。

K:表示键值对中键的类型。

V:表示键值对中值的类型。

T:表示其他所有的类型。

泛型类可以被继承,与普通类不同。

例如,下面代码定义了泛型类 ClassA,在 ClassADemo 测试类中构建两个实例 c1 和 c2,c1 两个属性类型为 String,c2 两个属性类型为 Integer,分别实现"+"运算。

```java
public class ClassADemo {
  public static void main(String[] args) {
    ClassA<String>c1=new ClassA<String>();
    c1.setS1("123");
    c1.setS2("456");
    c1.add();
    ClassA c2=new ClassA(123,456);
    //ClassA c2=new ClassA(new Integer(123),new Integer(456));
    c2.add();
  }
}
```

```
class ClassA<T>{
    private T s1;
    private T s2;
    public T getS1() {
        return s1;
    }
    public void setS1(T s1) {
        this.s1 = s1;
    }
    public T getS2() {
        return s2;
    }
    public void setS2(T s2) {
        this.s2 = s2;
    }
    public ClassA(){}
    public ClassA(T s1,T s2){
        this.s1=s1;
        this.s2=s2;
    }
    public void add(){
        if(s1.getClass().getName()==Integer.class.getName()){
            int l1=(int) s1;
            int l2=(int)s2;
            System.out.println("结果是:"+(l1+l2));
        }else if(s1.getClass().getName()==String.class.getName()){
            System.out.println("结果是:"+s1+s2);
        }else{
            System.out.println("类型是:"+s1.getClass().getName());
        }
    }
}
```

程序运行结果如图 11-1 所示。

说明：

(1) T 是类型的参数，可以通过变量传递类型。

(2) 上题中类型传递方式有两种：ClassA＜String＞ c1 = new ClassA＜String＞();或者 ClassA c2＝new ClassA(123,456);。

(3) s1.getClass().getName()＝＝Integer.class.getName()，是获取的全限定类名进行比较。

结果是：123456
结果是：579

图 11-1　程序运行结果

11.1.2　泛型类的子类及有界类型参数

泛型类可以被继承，继承时子类需要把类型传递给父类，语法格式如下：

```
class classA<T>{}
class classB<T>extends classA{}
```

Java 泛型的本质是类型的参数化，在实际应用中可以被任何需要的类型替代。但是在某些特定的情景下，参数的类型需要进行限制。Java 语言通过 extends 设置类型的范围，具体格式如下：

```
Class classA<T extends 类型 B>{}
```

extends 代表不是继承，是指定上界。T 类型是类型 B 或者是 B 的子类，T＜＝类型 B。若 T 有多个上界时使用 & 分割。

11.1.3 Collection

Java 中的集合又称为容器，主要的作用是存储对象。可以动态地将多个对象的引用放入容器中，其中不仅可以存储数量不等的对象的引用，还可以存储有映射关系的关联数组。集合框架主要由一组操作对象的接口组成，主要分为 Collection 和 Map 两大接口。这两大接口并不提供具体的实现，操作集合时，必须通过接口下的实现类进行。图 11-2 为集合框架的继承关系图，虚线箭头表示实现，实线箭头表示继承。

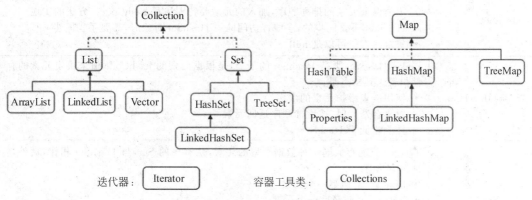

图 11-2　集合框架的继承关系图

Collection 接口是集合中的底层接口，提供了很多常用的方法，具体见表 11-1。

表 11-1　Collection 接口提供的常用方法

方　　法	功　　能
add(Object o)	增加元素
addAll(Collection c)	添加对象 c 中的所有元素
clear()	清空元素
contains(Object o)	是否包含指定元素
containsAll(Collection c)	是否包含集合 c 中的所有元素
iterator()	返回 Iterator 对象，用于遍历集合中的元素
remove(Object o)	移除元素
removeAll(Collection c)	相当于减集合 c
retainAll(Collection c)	相当于求与 c 的交集

续表

方法	功能
size()	返回元素个数
toArray()	把集合转换为一个数组

通过 Iterator 迭代器的 hasNext() 方法及 Next() 方法可以逐个访问 Collection 元素的方法。

11.1.4 Set 接口及其实现类

Set 是数学集合模型的抽象定义。Set 接口具有无序性和不重复性两大特点。Set 接口的方法是 Collection 中定义的通用方法。Set 集合使用 equals() 方法来判断元素是否相等。Set 接口常见的实现类如表 11-2 所示。

表 11-2 Set 接口常见的实现类

类 名	特 点 描 述
HashSet	HashSet 是 Set 接口的典型实现。特点是： • 不能保证元素的排列顺序，加入的元素要特别注意 hashCode() 方法的实现 • HashSet 不是同步的，多线程访问同一 HashSet 对象时，需要手工同步 • 集合元素值可以是 null
LinkedHashSet	LinkedHashSet 类是 HashSet 的子类，是根据元素的 hashCode 值来决定元素的存储位置。其特点是： • 使用链表维护元素的次序 • 对集合迭代时，按增加顺序返回元素 • 性能略低于 HashSet
TreeSet	TreeSet 只能存放同一种数据类型的元素，是有序的 Set，与 HashSet 相比，额外增加的方法有： • first()：返回第一个元素 • last()：返回最后一个元素 • lower(Object o)：返回指定元素之前的元素 • higher(Object o)：返回指定元素之后的元素 • subSet(fromElement, toElement)：返回子集合
EnumSet	EnumSet 类是专为枚举类设计的集合类，EnumSet 中的所有元素都必须是指定枚举类型的枚举值

11.1.5 List 接口及其实现类

List 接口主要用来存储有序的、可以重复的元素。List 接口在 Collection 接口中的通用方法中添加了一些有关索引的方法，具体见表 11-3。

表 11-3 List 接口中的索引方法

方法	功能
add(int index, Object o)	在指定位置插入元素
addAll(int index, Collection c)	在指定位置处添加集合 c
get(int index)	取得指定位置元素

续表

方　法	功　能
indexOf(Object o)	返回对象 o 在集合中第一次出现的位置
lastIndexOf(Object o)	返回指定元素在集合中最后一次出现的位置，若不存在返回 −1
remove(int index)	删除并返回指定位置的元素
set(int index, Object o)	替换指定位置元素
subList(int fromIndex, int endIndex)	返回子集合

实现接口 List 的类有 ArrayList、Vector 和 LinkedList。ArrayList 是最常用的 List 的实现类，ArrayList 是基于数组实现的线性表，是不安全的线程。Vector 是安全的线程，但性能比 ArrayList 低。LinkedList 类是基于链表实现的，可频繁进行插入和删除操作，性能较高。

11.1.6　Map 接口及其实现类

Map 接口和 Collection 接口是并列的，其实现类常用来保存 K-V 形式的数据。Map 接口的方法见表 11-4。

表 11-4　Map 接口的方法

方　法	功　能
clear()	从此映射中移除所有映射关系
containsKey(Object k)	如果此映射包含指定键的映射关系，则返回 true
containsValue(Object v)	如果此映射将一个或多个键映射到指定值，则返回 true
entrySet()	返回此映射中包含的映射关系的 Set 视图
equals(Object obj)	比较指定的对象与此映射是否相等
get(Object k)	返回指定键所映射的值；如果此映射不包含该键的映射关系，则返回 null
hashCode()	返回此映射的哈希码值
isEmpty()	如果此映射未包含键-值映射关系，则返回 true
keySet()	返回此映射中包含的键的 Set 视图
put(Object k, Object v)	将指定的值与此映射中的指定键关联
putAll(Map m)	从指定映射中将所有映射关系复制到此映射中
remove(Object k)	如果存在一个键的映射关系，则将其从此映射中移除
size()	返回此映射中的键-值映射关系数
values()	返回此映射中包含的值的 Collection 视图

Map 接口常见的实现类见表 11-5。

表 11-5　Map 接口常见的实现类

类　名	描　述
HashMap	HashMap 是 Map 中最常见的类，它根据键的 hashCode 值存储数据。大多数情况下可以直接定位到它的值，因此具有很快的访问速度，但遍历顺序不确定。HashMap 最多只允许一条记录的键为 null，允许多条记录的值为 null
HashTable	HashTable 是线程安全的比较古老的 Map 实现类

续表

类 名	描 述
LinkedHashMap	LinkedHashMap 是 HashMap 的一个子类,遍历效果较好,插入和删除操作效率较低
TreeMap	TreeMap 实现 SortedMap 接口,能够把它保存的记录根据键值排序,默认是按键值的升序排序,也可以指定排序的比较器

11.2 实例解析

11.2.1 实例1:List集合的基本使用

【实例要求】

编写程序练习 List 集合的基本使用。

(1) 创建一个只能容纳 String 对象名为 names 的 ArrayList 集合。

(2) 按顺序往集合中添加 5 个字符串对象:张三、李四、王五、马六、赵七。

(3) 对集合进行遍历,分别打印集合中的每个元素的位置与内容。

(4) 首先打印集合的大小,然后删除集合中的第 3 个元素,并显示删除元素的内容,然后再打印目前集合中第 3 个元素的内容,并再次打印集合的大小。如图 11-3 所示。

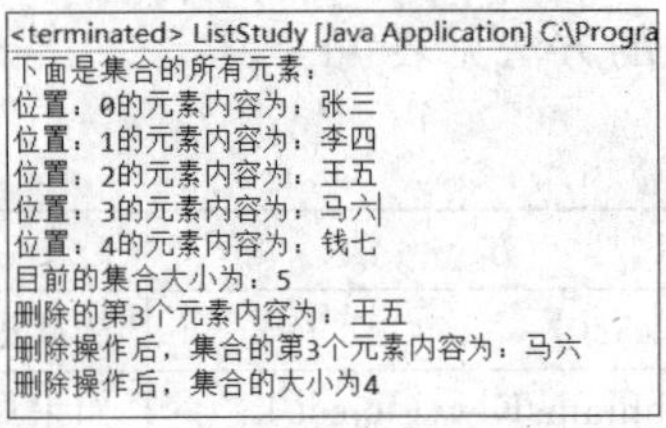

图 11-3 调试结果

【实例源代码】

新建源文件 shiyan1.java,源代码如下:

```java
gdlgxy.shiyan;
/* 功能描述:List 集合的基本使用 */
import java.util.ArrayList;
import java.util.List;
public class shiyan1 {
  public static void main (String[] args){
    List<String>names=new ArrayList <String>();
    System.out.println("下面是集合的所有元素:");
    names.add("张三");
    names.add("李四");
    names.add("王五");
    names.add("马六");
    names.add("钱七");
    for (int i=0;i<names.size();i++){
        System.out.println("位置:"+i+"的元素内容为:"+names.get(i));
    }
    System.out.println("目前的集合大小为:"+names.size());
```

```
        System.out.println("删除的第 3 个元素内容为:"+names.get(2));
        names.remove(2);
        System.out.println("删除操作后,集合的第 3 个元素内容为:"+names.get(2));
        System.out.println("删除操作后,集合的大小为"+names.size());
    }
}
```

【实例解析】

List 接口主要用来存储有序的、可以重复的元素,其又可以分为 ArrayList、LinkedList、Vector 三个具体的实现类。

为了存储有序元素,List 接口在 Collection 接口的基础上又增加了一些方法。这些新增方法的主要特点是可以通过索引对集合的有序元素进行操作。新增方法如下。

1. 增加(插入)元素

void add(int index，Object ele)：在指定的索引位置 index 处添加元素 ele。

boolean addAll(int index，Collection eles)：在指定的索引位置 index 处添加集合 eles。

2. 删除元素

Object remove(int index)：删除指定索引位置的元素。

3. 修改元素

Object set(int index，Object ele)：设置指定索引位置的元素为 ele。

4. 查询元素

(1) Object get(int index)：获取指定索引位置的元素。

(2) int indexOf(Object obj)：返回指定元素在集合中首次出现的位置,不存在则返回 -1。

(3) int lastIndexOf(Object obj)：返回指定元素在集合中最后出现的位置,不存在则返回 -1。

(4) List sublist(int fromIndex, int toIndex)：返回从 fromIndex 位置开始到 toIndex-1 位置结束的一个子 List。

上面这些新增方法可以用于 List 接口下的任意实现类。ArrayList 是 List 接口下最常用的实现类,其底层是以数组结构实现的。

注意：Vector 是 List 下一个古老的实现类,与 ArrayList 相比,Vector 是线程安全的,但是执行效率较低,不推荐使用。

11.2.2 实例 2：Map 集合的基本使用

【实例要求】

编写程序练习 Map 集合的基本使用。

(1) 创建一个只能容纳 String 对象的 person 的 HashMap 集合。

(2) 向集合中添加 5 个"键-值"对象：id→"1"、name→"张三"、sex→"男"、age→"25"、

love→"学习 Java"。

(3) 对集合进行遍历,分别打印集合中的每个元素的键与值。

(4) 首先打印集合的大小,然后删除集合中的键为 age 的元素,并显示删除元素的内容,并再次打印集合的大小。

调试结果如图 11-4 所示。

```
Problems  Javadoc  Declaration  Console
<terminated> shiyan2 [Java Application] C:\Program Files\Java\jre1.8.0
下面是集合的所有元素:
键: love-->值学习Java
键: age-->值25
键: sex-->值男
键: name-->值张三
键: id-->值1
目前集合的大小为: 5
删除的键age的内容为: null
删除操作后,集合的大小为5
```

图 11-4　调试结果

【实例源代码】

新建源文件 shiyan2.java,源代码如下:

```java
package gdlgxy.shiyan;
import java.util.HashMap;
import java.util.Iterator;
import java.util.Map;
public class shiyan2 {
    /* 功能描述:Map 集合的基本使用 */
    public static void main (String[] args){
        Map<String,String>person=new HashMap<String,String>();
        System.out.println("下面是集合的所有元素:");
        person.put("id","1");
        person.put("name","张三");
        person.put("sex","男");
        person.put("age","25");
        person.put("love","学习 Java");
        for(Iterator<java.util.Map.Entry<String, String>>iter=person.entrySet().iterator();iter.hasNext();){
            Map.Entry entry=(Map.Entry) iter.next();
            System.out.println("键:"+entry.getKey()+"-->值"+entry.getValue());
        }
        System.out.println("目前集合的大小为:"+person.size());
        System.out.println("删除的键 age 的内容为:"+person.get("age"));
        person.remove("age");
        System.out.println("删除操作后,集合的大小为"+person.size());
    }
}
```

【实例解析】

Map 中最常用的方法主要有添加、删除、查询及遍历。
- 添加、删除的操作方法如下。
    - Object put(Object key,Object value)：向 Map 中添加一个元素。
    - Object remove(Object key)：按照指定 key 删除 Entry。
    - void putAll(Map t)：将 Map t 中的所有元素添加到当前 Map 中。
    - void clear()：清空元素。
- 元素查询的操作方法如下。
    - Object get(Object key)：获取指定 key 的 value 值。若无 key,则返回 null 值。
    - boolean containsKey(Object key)：是否包含指定 key。
    - boolean containsValue(Object value)：是否包含指定 value。
    - int size()：返回集合长度。
    - boolean isEmpty()：查询是否为空。
    - boolean equals(Object obj)：判断元素是否相等。key 和 value 的值都要判断。
- 用于遍历 Map 的方法称为元视图操作法,主要有 3 个方法。
    - Set keySet()：遍历 key 集。
    - Collection values()：遍历 value 集。
    - Set entrySet()：遍历 key-value 对,即遍历 Entry。

11.3 上机实验

【实验目的】

- 掌握集合的概念、体系结构、分类及使用场景。
- 了解 Collection 接口及主要实现类(List、Set)。
- 掌握 List、Set 的使用方法。
- 了解 Map 接口及主要实现类(HashMap、TreeMap、HashTable)。
- 掌握 HashMap、TreeMap 的使用方法。

【实验要求】

创建一个 Worker 类,为 Worker 类添加相应的代码,使得 Worker 对象能正确放入 TreeSet 中,并编写相应的测试代码。

比较时,先比较工人年龄大小,年龄小的排在前面。如果两名工人年龄相同,则比较其收入,收入少的排前面。如果年龄和收入都相同,则根据字典顺序比较工人姓名。例如,有四名工人,基本信息如下：

姓名	年龄	工资
zhang3	18	1500
li4	18	1500

```
wang5    18    1600
zhao6    17    2000
```

放入 TreeSet 排序后结果为 zhao6 li4 zhang3 wang5。

实验运行效果如图 11-5 所示。

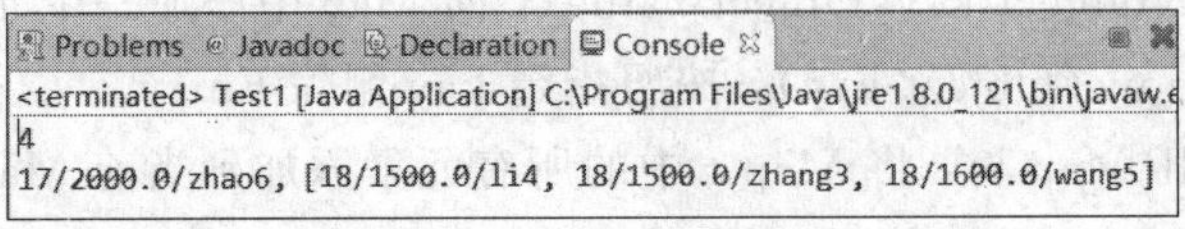

图 11-5 实验运行效果

【实验指导】

向 TreeSet 中添加自定义类的对象时，排序方式有两种：自然排序和定制排序。

（1）自然排序：要求必须在自定义类中实现 java.lang.Comparable 接口，并在其中重写 compareTo(Object obj) 方法，指明要按照哪个属性进行排序。

（2）定制排序方法的步骤如下：

① 创建一个 Comparator 接口的类对象。

② 重写 compare() 方法，在此方法中指明按照自定义类中的哪个属性进行排序。

③ 将此对象作为形参传递给 TreeSet 的构造器。

④ 在 TreeSet 中添加 compare 中比较的自定义类的元素即可。

【程序模板】

```java
import java.util.HashSet;
import Demo0.Worker;
public class CharacterTest1 {
    /* 功能描述:比较工人年龄大小,年龄小的排在前面。如果两个工人年龄相同,则比较其收入,
       收入少的排前面。如果年龄和收入都相同,则根据字典顺序比较工人姓名 */
    public static void main(String[] args) {
        //TODO Auto-generated method stub
        HashSet<Worker>hs=new HashSet<Worker>();
        Worker w1=new Worker("zhang3", 18, 1500);
        [代码1]    //增加 w2 "li4",18,1500
        [代码2]    //增加 w3 "wang5",18,1600
        [代码3]    //增加 w4 "zhao6",17,2000
        hs.add(w1);
        hs.add(w2);
        hs.add(w3);
        hs.add(w4);
        System.out.println(hs.size());
        System.out.println(hs);
    }
}
class Worker implements Comparable<Worker>{
```

```java
    String name;
    int age;
    double salary;
    public Worker(){}
    public Worker(String name, int age, double salary){
        this.name =name;
        this.age =age;
        this.salary =salary;
    }
    public int compareTo(Worker o) {
        //TODO Auto-generated method stub
        if(this.age!=o.age){
            return this.age-o.age;
        }
        else if(this.salary!=o.salary){
            //Integer integer1=new Integer(this.salary)
            return new Double(this.salary).compareTo(new Double(o.salary));
        }
        else if(this.name.equals(o.name)){
            return this.name.compareTo(o.name);
        }
        return 0;
    }
    public int hashCode() {
        final int prime =31;
        int result =1;
        result =prime * result +age;
        result =prime * result +((name ==null) 0 : name.hashCode());
        long temp;
        temp =Double.doubleToLongBits(salary);
        result =prime * result +(int) (temp ^ (temp >>> 32));
        return result;
    }
    public boolean equals(Object obj) {
        if (this ==obj)
            return true;
        if (obj ==null)
            return false;
        if (getClass() !=obj.getClass())
            return false;
        Worker other =(Worker) obj;
        if (age !=other.age)
            return false;
        if (name ==null) {
            if (other.name !=null)
                return false;
        } else if (!name.equals(other.name))
            return false;
        if (Double.doubleToLongBits (salary) != Double.doubleToLongBits (other.salary))
```

```
            return false;
        return true;
    }
    public String toString() {
        //TODO Auto-generated method stub
        return age+"/"+salary+"/"+name;
    }
}
```

11.4 拓展练习

【基础知识练习】

一、选择题

1. 集合 API 中 Set 接口的特点是(　　)。
   A. 不允许重复元素,元素有顺序   B. 允许重复元素,元素无顺序
   C. 允许重复元素,元素有顺序   D. 不允许重复元素,元素无顺序
2. List 接口的特点是(　　)。
   A. 不允许重复元素,元素有顺序   B. 不允许重复元素,元素无顺序
   C. 允许重复元素,元素有顺序   D. 允许重复元素,元素无顺序
3. 执行以下程序后的结果是(　　)。

```
public class Demo {
    public static void main(String[] args) {
        List a1 =new ArrayList();
        a1.add("1");
        a1.add("2");
        a1.add("2");
        a1.add("3");
        System.out.println(a1);
    }
}
```

   A. [1, 2, 3]   B. [1, 2, 3, 3]   C. [1, 2, 2, 3]   D. [2, 1, 3, 2]

二、判断题

1. List 接口中的内容是不能重复的。   (　　)
2. TreeSet 是排序类。   (　　)
3. Set 接口的内容可以使用 Enumeration 接口进行输出。   (　　)
4. Map 接口的内容可以使用 ListIterator 接口进行输出。   (　　)

【上机练习】

已知有 16 支男子足球队参加 2008 年北京奥运会。编写一个程序,把这 16 支球队随机

分为4个组,如图11-6所示。

（1）参赛球队有：澳大利亚、阿根廷、科特迪瓦、尼日利亚、荷兰、日本、美国、塞尔维亚、喀麦隆、中国、洪都拉斯、新西兰、巴西、意大利、比利时、韩国。

（2）使用 Math.random 产生随机数。

尼日利亚	新西兰	巴西	喀麦隆
日本	美国	中国	韩国
塞尔维亚	比利时	洪都拉斯	意大利
科特迪瓦	阿根廷	澳大利亚	荷兰

图 11-6 实验运行效果

GUI 编程

12.1 知识提炼

12.1.1 GUI 概述

图形用户界面(graphic user interface,GUI)是为方便用户使用而设计的窗口界面,在图形用户界面中用户可以看到什么就操作什么,取代了在字符方式下知道是什么后才能操作什么的方式。组件(component)是构成 GUI 的基本要素,通过对不同事件的响应来完成和用户的交互或组件之间的交互。组件主要通过 java.awt 包中的重量级类和 javax.swing 包中的轻量级类实现,本章主要介绍使用 Swing 组件。组件一般作为一个对象放置在容器(container)内。

12.1.2 容器组件

在 Swing 组件中,使用较多的容器类有 JFrame、JPanel 和 JDialog 等。JFrame 是带有标题栏和边界的窗口,而且允许调整大小。另外,用户还可以为窗口附加一个菜单栏。当程序窗口需要图表化或者需要包含菜单栏时,用户在程序设计过程中则可以选择使用窗口组件。JFrame 构造一个窗口后,可以用 add()方法来给窗口添加面板 JPanel、基本组件,也可以设置菜单,JFrame 窗口是界面底层容器。

1. JFrame 的构造方法

(1) JFrame():创建无标题窗口。

(2) JFrame(String s):创建标题名是字符串 s 的窗口。

2. JFrame 的常用方法

(1) void setTitle(String title):设置窗口标题文本。

(2) void setSize():设置窗口的大小。

(3) void setLayout(LayoutManager manager):设置窗口的布局。

(4) void setJMenuBar(JMenuBar menubar):设置窗口的菜单栏。

(5) void setVisible(boolean b):设置窗口的可见性。

(6) void setLocation(x,y):设置窗口在屏幕的位置。

(7) void setDefaultCloseOperation(int operation):设置单击窗口右上角的关闭按钮所执行的操作。

12.1.3 基本组件

Swing 包中的基本组件有：JLabel 标签、JButton 按钮、TextField 文本框、JPasswordField 密码框、JRadioButton 单选按钮、JCheckBox 复选框、JComboBox 下拉列表框、JList 列表框和 JTextArea 文本区。

12.1.4 布局管理器

布局管理器是控制 GUI 布局。基本的布局管理器有：流式布局、边界式布局、网格式布局、卡片式布局、null 布局，见表 12-1。如果窗体布局比较复杂且组件较多就要使用容器的嵌套，每一种容器都设置一种布局管理器。当容器大小改变时，布局管理器能自动调整组件的排列。

表 12-1 布局管理器对比表

名 称	功 能
FlowLayout	该布局称为流式布局管理器，是从左到右，中间放置，一行放不下就换到另外一行。一行能放置多少组件取决于窗口的宽度。默认组件是居中对齐，可以通过 FlowLayout(intalign) 函数来指定对齐方式，默认情况下是居中（FlowLayout. CENTER）
BorderLayout	这种布局管理器分为东、南、西、北、中心五个方位。北和南的组件可以在水平方向上拉伸；而东和西的组件可以在垂直方向上拉伸；中心的组件可以在水平和垂直方向上同时拉伸，从而填充所有剩余空间。在使用 BorderLayout 时，如果容器的大小发生变化，其变化规律为：组件的相对位置不变，大小发生变化
CardLayout	这种布局管理器能够帮助用户处理两个以至于更多的成员共享同一显示空间，它把容器分成许多层，每层的显示空间占据整个容器大小，但是每层只允许放置一个组件，当然每层都可以利用 Panel 来实现复杂的用户界面
GridLayout	这种布局是网格式的布局，窗口改变的时候，组件的大小也会随之改变。每个单元格的大小一样，而且放置组件时，只能按照从左到右、由上到下的顺序填充，用户不能任意放置组件
GridBagLayout	可以完成复杂的布局，而且 IDE 对它有足够的支持，是一个很强大的 Layout。不过它过于复杂，在此布局中，组件大小不必相同

12.1.5 事件处理

掌握事件处理机制、事件常用的监听器及监听方法和常用的事件处理。常用的事件类、监听器接口/适配器及方法以及触发事件的用户操作等如表 12-2 所示。

表 12-2 事件类、监听器接口/适配器及方法以及触发事件的用户操作

事 件 类	监听器接口/适配器	方法以及触发事件的用户操作
ActionEvent（动作事件）	ActionListener/无	actionPerformed(ActionEvent e) 单击按钮、菜单、按钮上按空格键、菜单上按 Enter 键、单选按钮与复选框和下拉组合框做出选择等
ItemEvent（选项事件）	ItemListener/无	itemStateChanged(ItemEvent e) 改变单选按钮、复选框、下拉组合框等选择

续表

事 件 类	监听器接口/适配器	方法以及触发事件的用户操作
TextEvent （文本事件）	TextListener/ 无	textValueChanged(TextEvent e)改变 java.awt 包中文本组件 TextField 和 TextArea 的内容
ComponentEvent （组件事件）	ComponentListener/ ComponentAdapter	componentHidden(ComponentEvent e)隐藏组件 componentMoved(ComponentEvent e)移动组件 componentResized(ComponentEvent e)改变组件大小 componentShown(ComponentEvent e)显示组件
FocusEvent （焦点事件）	FocusListener/ FocusAdapter	focusGained(FocusEvent e)组件获得键盘焦点 focusLost(FocusEvent e)组件失去键盘焦点
ContainerEvent （容器事件）	ContainerListener/ ContainerAdapter	componentAdded(ContainerEvent e)添加组件到容器 componentRemoved(ContainerEvent e)移除容器组件
WindowEvent （窗口事件）	WindowListener/ WindowAdapter	windowOpened(WindowEvent e)打开窗口 windowActivated(WindowEvent e)激活窗口 windowClosed(WindowEvent e)关闭窗口后 windowClosing(WindowEvent e)正在关闭窗口 windowDeactivated(WindowEvent e)停用窗口 windowDeiconified(WindowEvent e)取消最小化窗口 windowIconified(WindowEvent e)最小化后窗口
	WindowStateListener	void windowStateChanged(WindowEvent)窗口状态改变调用
KeyEvent （键盘事件）	KeyListener/ KeyAdapter	keyPressed(KeyEvent e)按下键盘按键 keyReleased(KeyEvent e)释放按键 keyTyped(KeyEvent e)敲击按键
MouseEvent （鼠标事件）	MouseListener/ MouseAdapter	mouseClicked(MouseEvent e)单击（按下并释放）鼠标 mouseEntered(MouseEvent e)鼠标进入组件 mouseExited(MouseEvent e)鼠标离开组件 mousePressed(MouseEvent e)按下鼠标 mouseReleased(MouseEvent e)释放鼠标
ListSelectionEvent （列表选择事件）	ListSelectionListener/ 无	valueChanged(ListSelectionEvent e)更改列表选项

12.1.6 菜单、其他组件

学习 JMenuBar 菜单栏、JMenu 菜单、JMenuItem 菜单项、JPopupMenu 弹出菜单和 JCheckBoxMenuItem 复选菜单的方法和使用，以及如何创建标准菜单系统。

学习 JColorChooser 颜色选择器、JSlider 滑动条、JProgressBar 进度条、JToolBar 工具条、JTabbedPane 选项卡、JSplitPane 分隔窗格、JTable 数据表格和 JTree 树组件的方法和使用。

12.2 实例解析

12.2.1 实例1：JList列表

【实例要求】

使用JList编写列表框程序：可以同时将多个选项信息以列表的方式提供给客户，第一个列表框中设置一次只可以选择一个选项，第二个列表框中设置一次可以选择多个选项，如图12-1所示。

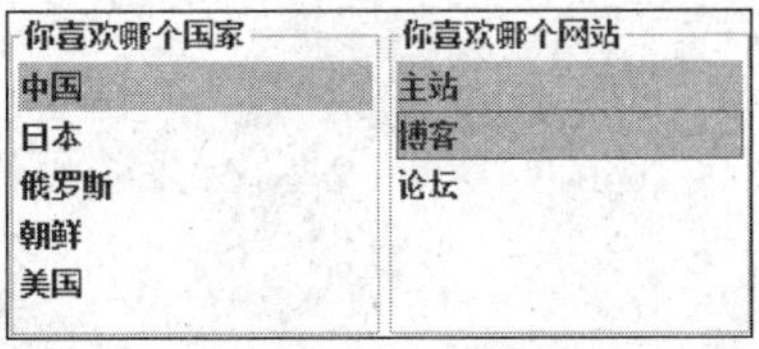

图12-1 列表框程序

【实例源代码】

新建源文件Shiyan1.java，源代码如下：

```java
package gdlgxy.shiyan;
    /* 功能描述:JList列表 */
import java.awt.Container;
import java.awt.GridLayout;
import java.awt.event.WindowAdapter;
import java.awt.event.WindowEvent;
import java.util.Vector;
import javax.swing.BorderFactory;
import javax.swing.JFrame;
import javax.swing.JList;
import javax.swing.ListSelectionModel;
class MyList {
  private JFrame frame =new JFrame("Beyole");
  private Container container =frame.getContentPane();
  private JList list1 =null;             //定义列表框
  private JList list2 =null;             //定义列表框
  public MyList() {
      this.frame.setLayout(new GridLayout(1, 2));
      String nation[] ={ "中国", "日本", "俄罗斯", "朝鲜", "美国" };
      Vector<String>vector =new Vector<String>();
      vector.add("主站");
      vector.add("博客");
      vector.add("论坛");
      this.list1 =new JList(nation);
      this.list2 =new JList(vector);
      list1.setBorder(BorderFactory.createTitledBorder("你喜欢哪个国家"));
      list2.setBorder(BorderFactory.createTitledBorder("你喜欢哪个网站"));
      list1.setSelectionMode(ListSelectionModel.SINGLE_SELECTION);
      list2.setSelectionMode(ListSelectionModel.MULTIPLE_INTERVAL_SELECTION);
      container.add(this.list1);
```

```
            container.add(this.list2);
            this.frame.setSize(330, 180);
            this.frame.setVisible(true);
            this.frame.addWindowListener(new WindowAdapter() {
                public void windowClosing(WindowEvent arg0) {
                    System.exit(1);
                }
            });
        }
    }
    public class shiyan1 {
      public static void main(String[] args) {
          new MyList();
      }
    }
```

【实例解析】

JList 组件一次可以选择一项或多项。选择多项时可以是连续区间选择(按住 Shift 键进行选择),也可以是不连续的选择(按住 Ctrl 键进行选择)。

(1) JList 的构造方法。

① JList():使用空的模式构造列表框。

② JList(Object[] listData):使用指定的数组构造列表框。

③ JList(VectorlistData):使用包含元素的向量构造列表框。

(2) JList 的常用方法。

① int getSelectedIndex():返回选中的索引选项,如果选中多个,则返回最小的一个索引。

② int[]getSelectedIndices():返回选中的索引选项,这些索引组成一个数组,升序排序。

③ boolean isSelectionEmpty():判断是否为空选择。

④ void setListData(Object[] listData):设置列表框选项为给定的数组元素。

⑤ void setVisibleRowCount(int x):设置列表框的可见选项行数。

> **注意**:以上程序中第一个 JList 通过字符串数组设置列表的内容,并通过 setSelectionMode()方法设置一次性可以选择的一个选项,第二个 JList 通过 Vector 设置列表的内容,并通过 setSelectionMode()方法设置一次性可以选择的多个选项。

(3) 养成良好的编程习惯,如缩进、写注释等,掌握单行注释和多行注释表示方法。

12.2.2 实例2:求三角形面积

【实例要求】

在文本框中输入三条边,计算三角形面积。若不能构成三角形或者是输入错误,就提示错误。如果可以计算三角形面积,就通过三条边计算三角形面积。运行结果如图 12-2 所示。

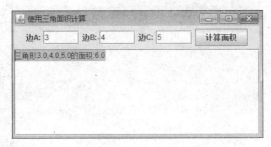

图 12-2　三角形面积计算运行界面

【实例源代码】

新建源文件 Triangle.java，源代码如下：

```java
public class Triangle {
    /* 功能描述：三角形类，判断是否能构成三角形、计算三角形面积 */
    double sideA, sideB, sideC, area;
    boolean isTriangle;
    public String getArea(){
        if(isTriangle){
            double p=(sideA+sideB+sideC)/2.0;
            area=Math.sqrt(p*(p-sideA)*(p-sideB)*(p-sideC));
            return String.valueOf(area);
        }
        else {
            return "无法计算面积";
        }
    }

    public void setA(double a){
        sideA=a;
        if(sideA+sideB>sideC&&sideA+sideC>sideB&&sideC+sideB>sideA)
            isTriangle=true;
        else
            isTriangle=false;
    }

    public void setB(double b){
        sideB=b;
        if(sideA+sideB>sideC&&sideA+sideC>sideB&&sideC+sideB>sideA)
            isTriangle=true;
        else
            isTriangle=false;
    }

    public void setC(double c){
        sideC=c;
```

```
        if(sideA+sideB>sideC&&sideA+sideC>sideB&&sideC+sideB>sideA)
            isTriangle=true;
        else
            isTriangle=false;
    }
}
```

新建源文件 WindowTriangle.java，源代码如下：

```
import java.awt.*;
import java.awt.event.*;
import javax.swing.*;
public class WindowTriangle extends JFrame implements ActionListener{
    /* 功能描述:窗口类,构建窗口 */
  Triangle triangle;
  JTextField textA,textB,textC;
  JTextArea showArea;
  JButton controlButton;
  WindowTriangle(){
      init();
      setVisible(true);
      setDefaultCloseOperation(JFrame.EXIT_ON_CLOSE);
  }

    void init(){
        triangle=new Triangle();
        textA=new JTextField(5);
        textB=new JTextField(5);
        textC=new JTextField(5);
        showArea=new JTextArea();
        controlButton=new JButton("计算面积");
        JPanel pNorth=new JPanel();
        pNorth.add(new JLabel("边 A:"));
        pNorth.add(textA);
        pNorth.add(new JLabel("边 B:"));
        pNorth.add(textB);
        pNorth.add(new JLabel("边 C:"));
        pNorth.add(textC);
        pNorth.add(controlButton);
        controlButton.addActionListener(this);
        add(pNorth,BorderLayout.NORTH);
        add(new JScrollPane(showArea),BorderLayout.CENTER);
    }
    public void actionPerformed(ActionEvent e){
        try{
            double a=Double.parseDouble(textA.getText().trim());
            double b=Double.parseDouble(textB.getText().trim());
            double c=Double.parseDouble(textC.getText().trim());
```

```
                triangle.setA(a);
                triangle.setB(b);
                triangle.setC(c);
                String area=triangle.getArea();
                showArea.append("三角形"+a+","+b+","+c+"的面积:");
                showArea.append(area+"\n");
            }
            catch (Exception ex){
                showArea.append("\n"+ex+"\n");
            }
        }
    }
}
```

新建源文件 Test.java,源代码如下:

```
public class Test {
    /* 功能描述:测试类,实例化窗口并显示窗口 */
    public static void main(String args[]){
        WindowTriangle win=new WindowTriangle();
        win.setTitle("使用三角形面积计算");
        win.setBounds(200,200,450,240);
    }
}
```

【实例解析】

(1) 程序的结构是基于 MVC 模型设计,MVC 是模型(model)、视图(view)、控制器(controller)的缩写,是一种软件设计典范。它是用一种业务逻辑、数据与界面显示分离的方法来组织代码,将众多的业务逻辑聚集到一个部件里面,在需要改进和个性化定制界面及用户交互的同时,不需要重新编写业务逻辑,达到减少编码的时间。

MVC 开始是存在于桌面程序中的,M 是指业务模型,V 是指用户界面,C 则是指控制器。

(2) init()方法是初始化方法。

(3) Triangle triangle;是在 WindowTriangle 声明三角形类。

(4) WindowTriangle extends JFrame implements ActionListener 是指类 WindowTriangle 继承 Java 的 JFrame 类,JFrame 是 Java 的窗体类,继承它可以重写它的一些方法达到更方便编程的作用。implements ActionListener 实现 ActionListener 接口为动作监听接口,是 Java swing 监听窗体动作的一个接口。

(5) 本程序中使用的三角形面积计算公式是海伦公式。知道三条边 a、b、c 求面积 S。公式如下:

$$S=\sqrt{p(p-a)(p-b)(p-c)}$$

其中 p 是周长的一半:

$$p = \frac{a+b+c}{2}$$

(6) getText().trim()的作用是在获得的文本中除去空格,trim()的作用是去掉字符串左、右的空格。

12.3 上机实验

【实验目的】

- 了解事件处理机制。
- 掌握常见的事件处理。
- 可以编写并运行一个事件处理程序。

12.3.1 实验1：对键盘每个操作的监控

【实验要求】

设计一个程序,实现针对键盘的每个操作的监控。
实验运行结果如图12-3所示。

【实验指导】

键盘是一种最常用的输入工具。在java Swing编程过程中,经常需要处理键盘事件,例如处理快捷键等。KeyListener的主要方法如表12-3所示。

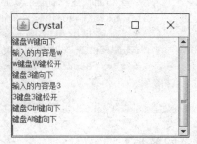

图12-3 实验运行结果

表12-3 KeyListener的主要方法

方　　法	功　　能
void keyPressed(KeyEvent)	按下按键
void keyReleased(KeyEvent)	释放按键
void keyTyped(KeyEvent)	敲击按键

【程序模板】

```java
import java.awt.BorderLayout;
import javax.swing.JPanel;
import javax.swing.JFrame;
import java.awt.Rectangle;
import javax.swing.JLabel;
import javax.swing.SwingConstants;
import java.awt.Font;
import javax.swing.JTextArea;
import javax.swing.JScrollBar;
import javax.swing.JScrollPane;
```

```
import javax.swing.JComboBox;
import javax.swing.JTextField;
import javax.swing.JButton;
import java.util.Date;
class MyKeyHandle extends JFrame implements KeyListener {
    private JTextArea text =new JTextArea();
    public MyKeyHandle() {
    /* 功能描述:对键盘的每个操作的监控 */
        super.setTitle("Crystal");
        [代码 1]   //加入滚动条
        pane.setBounds(5, 5, 300, 200);
        super.add(pane);                    //向窗体加入组件
        text.addKeyListener(this);          //加入 key 监听
        super.setSize(310, 210);            //设置窗体
        super.setVisible(true);             //显示窗体
        super.addWindowListener(new WindowAdapter() {
            public void windowClosing(WindowEvent arg0) {
                System.exit(1);             //系统退出
            }
        });
    }
        public void keyPressed(KeyEvent e) {
            text.append("键盘" +KeyEvent.getKeyText(e.getKeyCode()) +"键向下\n");
            //按下按键
        }
        public void keyReleased(KeyEvent e) {
            [代码 2]   //释放按键
        }
        public void keyTyped(KeyEvent e) {
            [代码 3]   //敲击按键
        }
}
public class CharacterTest1 {
    public static void main(String[] args) {
        new MyKeyHandle();
    }
}
```

12.3.2 实验 2:设计留言板

【实验要求】

设计一个基于 Swing 组件的简单留言板,但不要求实现网络功能,只要在本机上输入并显示留言就可以。留言板功能如下。

功能 1:输入留言内容后按下"发帖"按钮,在留言板上留言内容就显示出来了,这是留言板最基本的功能。

功能 2:留言内容需要重新编辑时,按下"删帖"按钮,就可以重新输入留言内容了。

功能 3：按"从头浏览"按钮，光标就会回到留言内容最开始的地方。
功能 4：按"从尾浏览"按钮，光标就会回到留言内容末尾的地方。
实验运行结果如图 12-4 所示。

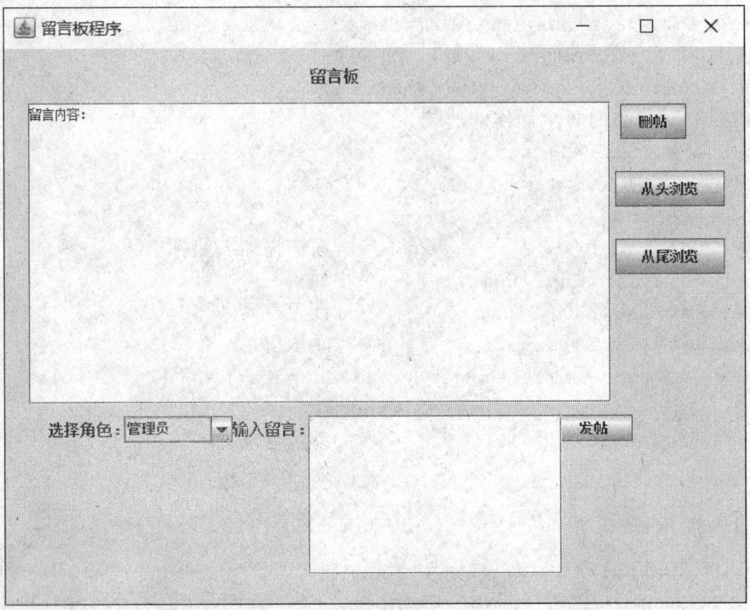

图 12-4　实验运行结果

【实验指导】

留言板上方为输出留言的文本区(JTextArea)，该文本区不可编辑。留言内容可能很长，所以在文本区应放一个滚动窗格(JScrollPane)。留言板的下方是输入留言的区域，输入姓名使用文本框(JTextField)，输入留言内容使用文本区(JTextArea)。

【程序模板】

```java
import java.awt.BorderLayout;
import javax.swing.JPanel;
import javax.swing.JFrame;
import java.awt.Rectangle;
import javax.swing.JLabel;
import javax.swing.SwingConstants;
import java.awt.Font;
import javax.swing.JTextArea;
import javax.swing.JScrollBar;
import javax.swing.JScrollPane;
import javax.swing.JComboBox;
import javax.swing.JTextField;
import javax.swing.JButton;
import java.util.Date;
```

```java
public class CharacterTest2 extends JFrame {
/* 功能描述:在本机上输入并显示留言 */
    private static final long serialVersionUID =1L;
    private JPanel jContentPane =null;
    private JLabel jLabel =null;
    private JTextArea jTextArea =null;
    private JScrollPane jScrollPane =null;
    private JLabel jLabel1 =null;
    private JComboBox jComboBox =null;
    private JLabel jLabel2 =null;
    private JTextField jTextField =null;
    private JButton jButton =null;
    private JButton jButton1 =null;
    private JButton jButton2 =null;
    private JButton jButton3 =null;
    /**
     * This is the default constructor
     */
    public CharacterTest2() {
        super();
        initialize();
    }
    /**
     * This method initializes this
     *
     */
    private void initialize() {
        this.setContentPane(getJContentPane());
        this.setTitle("留言板程序");
        this.setBounds(new Rectangle(0, 0, 640, 680));
        this.setVisible(true);
    }
    /**
     * This method initializes jContentPane
     *
     */
    private JPanel getJContentPane() {
        if (jContentPane ==null) {
            jLabel2 =new JLabel();
            jLabel2.setBounds(new Rectangle(210, 407, 90, 24));
            jLabel2.setFont(new Font("Dialog", Font.BOLD, 14));
[代码1]      //定义此组件要显示的单行文本为"输入留言"
            jLabel1 =new JLabel();
            jLabel1.setBounds(new Rectangle(40, 407, 90, 24));
            jLabel1.setFont(new Font("Dialog", Font.BOLD, 14));
[代码2]      //定义此组件要显示的单行文本为"选择角色"
            jLabel =new JLabel();
            jLabel.setBounds(new Rectangle(283, 15, 45, 20));
```

```java
                        jLabel.setHorizontalAlignment(SwingConstants.CENTER);
                        jLabel.setFont(new Font("Dialog", Font.BOLD, 14));
            [代码3]      //定义此组件要显示的单行文本为"留言板"
                        jContentPane = new JPanel();
                        jContentPane.setLayout(null);
                        jContentPane.add(jLabel, null);
                        jContentPane.add(getJScrollPane(), null);
                        jContentPane.add(jLabel1, null);
                        jContentPane.add(getJComboBox(), null);
                        jContentPane.add(jLabel2, null);
                        jContentPane.add(getJTextField(), null);
                        jContentPane.add(getJButton(), null);
                        jContentPane.add(getJButton1(), null);
                        jContentPane.add(getJButton2(), null);
                        jContentPane.add(getJButton3(), null);
            }
        return jContentPane;
    }
    /**
     * This method initializes jTextArea
     *
     */
    private JTextArea getJTextArea() {
        if (jTextArea == null) {
            jTextArea = new JTextArea();
            [代码4]      //构造一个新的文本区"留言内容"
            jTextArea.setEditable(false);
        }
            return jTextArea;
    }
    /**
     * This method initializes jScrollPane
     *
     */
    private JScrollPane getJScrollPane() {
        if (jScrollPane == null) {
            jScrollPane = new JScrollPane();
            jScrollPane.setBounds(new Rectangle(22, 49, 534, 347));
            jScrollPane.setViewportView(getJTextArea());
        }
        return jScrollPane;
    }
    /**
     * This method initializes jComboBox
     *
     */
    private JComboBox getJComboBox() {
        if (jComboBox == null) {
            [代码5]      //构造一个下拉组合框
```

```java
                jComboBox.setBounds(new Rectangle(110, 407, 100, 24));
                String[] mycbox={"管理员","博主","游客"};
                jComboBox.addItem(mycbox[0]);
                jComboBox.addItem(mycbox[1]);
                jComboBox.addItem(mycbox[2]);
            }
            return jComboBox;
        }
        /**
         * This method initializes jTextField
         *
         */
        private JTextField getJTextField() {
            if (jTextField ==null) {
                jTextField =new JTextField();
                jTextField.setBounds(new Rectangle(280, 407, 231, 200));
            }
            return jTextField;
        }
        /**
         * This method initializes jButton
         *
         */
        private JButton getJButton() {
            if (jButton ==null) {
                jButton =new JButton();
                jButton.setBounds(new Rectangle(505, 407, 70, 24));
                [代码 6]   //设置当前按钮名称为"发帖"
                jButton.addActionListener(new java.awt.event.ActionListener() {
                    public void actionPerformed(java.awt.event.ActionEvent e) {
                    Date date =new Date();
                    jTextArea.setText(jTextArea.getText() +"\r\n"+date.toString()+
"  "+jComboBox.getSelectedItem().toString()+"    留言:"+jTextField.getText());
                    }
                });
            }
            return jButton;
        }
        /**
         * This method initializes jButton1
         */
        private JButton getJButton1() {
            if (jButton1 ==null) {
                jButton1 =new JButton();
                jButton1.setBounds(new Rectangle(565, 51, 60, 32));
                jButton1.setText("删帖");
                jButton1.addActionListener(new java.awt.event.ActionListener() {
                    public void actionPerformed(java.awt.event.ActionEvent e) {
                    //jTextArea.setText("留言内容:");
```

```java
                    //if(jComboBox.getSelectedItem().toString() == "管理员" ||
jComboBox.getSelectedItem().toString() =="博主"){
                }
            });
        }
        return jButton1;
    }
    /**
     * This method initializes jButton2
     *
     */
    private JButton getJButton2() {
        if (jButton2 ==null) {
            jButton2 =new JButton();
            jButton2.setBounds(new Rectangle(560, 112, 100, 32));
            jButton2.setText("从头浏览");
            jButton2.addActionListener(new java.awt.event.ActionListener() {
                public void actionPerformed(java.awt.event.ActionEvent e) {
                    jTextArea.setCaretPosition(0);
                }
            });
        }
        return jButton2;
    }
    /**
     * This method initializes jButton3
     *
     * @return javax.swing.JButton
     */
    private JButton getJButton3() {
        if (jButton3 ==null) {
            jButton3 =new JButton();
            jButton3.setBounds(new Rectangle(560, 173, 100, 32));
            jButton3.setText("从尾浏览");
            jButton3.addActionListener(new java.awt.event.ActionListener() {
                public void actionPerformed(java.awt.event.ActionEvent e) {
                    jTextArea.setCaretPosition((int)jTextArea.getText().length());
                }
            });
        }
        return jButton3;
    }
    public static void main(String args[]){
        new CharacterTest2();
    }
}
```

12.4 拓展练习

【基础知识练习】

一、选择题

1. 下列有关 Swing 的叙述，错误的选项是（　　）。
　　A. Swing 是 Java 基础类（JFC）的组成部分
　　B. Swing 是可用来构建 GUI 的程序包
　　C. Swing 是 AWT 图形工具包的替代技术
　　D. Java 基础类（JFC）是 Swing 的组成部分

2. Swing GUI 通常由哪几类元素组成（　　）（选三项）。
　　A. GUI 容器　　　　　　　　　　B. GUI 组件
　　C. 布局管理器　　　　　　　　　D. GUI 事件侦听器

3. JTextField 类提供的 GUI 功能是（　　）。
　　A. 文本区域　　B. 按钮　　C. 文本字段　　D. 菜单

二、填空题

1. Java 编程语言是一种跨平台的编程语言，在编写图形用户界面方面，也要支持_____功能。

2. java.awt 包提供了基本的 Java 程序的 GUI 设计工具，主要包括下述三个概念，它们分别是_____、_____和_____。

3. BorderLayout 是_____、_____、_____和_____的默认布局策略。

【上机练习】

使用 Java 语言制作一个身高体重指数计算器。功能可选择三个标准：中国、亚洲、世界卫生组织（WHO）标准，计算公式：BMI＝weight/(height * height)。实验运行结果如图 12-5 所示。

图 12-5　实验运行结果

第13章

JDBC 编 程

13.1 知识提炼

13.1.1 数据库管理系统

数据库管理系统是一种操纵和管理数据库的大型软件,是用于建立、使用和维护数据库的,简称 DBMS。它对数据库进行统一的管理和控制,以保证数据库的安全性和完整性。用户通过 DBMS 访问数据库中的数据,数据库管理员通过 DBMS 进行数据库的维护工作。它提供多种功能,可使多个应用程序和用户用不同的方法在同时刻或不同时刻去建立、修改和询问数据库。目前有许多数据库产品,如 Oracle、Sybase、Informix、Microsoft SQL Server、Microsoft Access 等各以自己特有的功能创建数据库,并可以在该数据库下建立若干个表。表是关系型数据库的基本单位,可以使用多种数据库产品来创建表。

13.1.2 JDBC 的概念

JDBC 是一套类集,可以用来开发使用 Java 的客户/服务器数据库应用程序。它支持 SQL 工业标准,使得开发商可以使用各类数据库格式而无须知道低层数据库的具体细节。JDBC 有两个重要的组件:驱动程序管理器和 JDBC-ODBC 桥。

1. 驱动程序管理器

Java 应用程序的平台与数据库无关性是通过驱动程序管理器实现的,在开发应用中,我们需要为每一种数据库格式使用不同的 JDBC 驱动程序,有时甚至需要为同一个格式的不同版本使用多个不同的驱动程序,这些驱动程序的选择由驱动程序管理器完成。

2. JDBC-ODBC 桥

因 ODBC 驱动程序由数据库厂商提供,JDBC 驱动程序不能直接作为 ODBC 的驱动程序,需使用数据厂商提供的 JDBC-ODBC 桥建立与数据库的联系,通过 JDBC-ODBC 桥可将 JDBC 的方法调用转换成 ODBC 的函数调用,从而实现 JDBC 驱动程序被用作 ODBC 的驱动程序。

13.1.3 JDBC API

Java 采用 JDBC 实现对数据库的连接访问。JDBC API 主要是在 java.sql 包中定义的类和方法。其中定义了 3 个重要接口。

(1) Connection 接口:代表与数据库的连接,通过 Connection 接口提供的 getMetaData()

方法可获取所连接的数据库的有关描述信息。

（2）Statement 接口：用来执行 SQL 语句并返回结果记录集。

（3）ResultSet 接口：SQL 语句执行后的结果记录集。

使用 JDBC 进行数据库操作的步骤是：加载 JDBC 驱动器，定义供连接的 URL，建立连接，建立 Statement 对象，执行操作，处理结果，关闭连接。

13.1.4　JDBC 数据库连接的基本步骤

Java 语言操作各种数据库时所使用的程序模板都是一样的，即

```
import java.sql.*;
//该程序应用 JDBC-ODBC 桥与数据库建立联系
public class JDBCDatabaseDemo{
    public static void main(String[]args) throws Exception{String url="jdbc:odbc:northwind";
    Connection conn; //建立连接类
    //告诉程序使用 JBDC-OBDC 桥创建数据库连接
    Class.forName("sun.jdbc.odbc.JdbcOdbcDriver");
        //使用 getConnection()方法建立连接,第一个参数定义用户名,第二个参数定义密码
    con=DriverManager.getConnection(url, "sa","");}
}
```

有关数据库操作的主要类包是 java.sql，使用 Java 语言与数据库建立连接的基本步骤如下。

（1）加载数据库驱动程序。

Java 程序要与 DBMS（数据库管理系统）建立连接，首先需要装载驱动程序。装载驱动程序只需要非常简单的一行代码。

```
Class.forName ("sun.jdbc.odbc.JdbcOdbcDriver");
```

注意：Class.forName ("sun.jdbc.odbc.JdbcOdbcDriver")；语句中的大小写。

这里不需要创建一个驱动程序类的实例，而是通过 DriverManager 登记它，因为调用 Class.forName 将自动加载驱动程序类。

（2）数据库建立连接。

与数据库建立连接的一般方法如下：

```
Connection con = DriverManager.getConnection (String url, String user, string password)
```

其中，参数 url 的编写方法如下：

```
jdbc:<protocol>:<DatabaseName>
```

protocol 指 jbdc 连接数据库的机制。

DatabaseName 指数据库的名字。
示例如下：

```
jdbc:odbc;FaqAccess
dbc:mysql://localhost:3306/faq
jdbc:oracle:thin:bemyfriend:1521:ORCL.WORLD
```

其中，参数 user、password 分别是连接数据库时数据库本身的用户名和口令。

（3）创建 JDBC Statements 对象。

Statement 对象用于把 SQL 语句发送到 DBMS。用 Connection 对象来创建 Statement 对象的实例，例如：

```
statement stmt = con.createStatement();
```

（4）执行 SQL 语句，并获得结果。

若 SQL 语句作为 Statement 对象的 executeUpdate()或 executeQuery()方法的参数，则执行 SQL 语句。对于 SELECT 语句来说，可以使用 executeQuery。要创建或修改表的语句，使用的方法则是 executeUpdate。

通过 executeQuery()方法得到的是一个结果集（ResultSet），通过 ResultSet 的 next()方法可定位到不同的记录，需取得字段可以用 getInt()及 getString()等方法。

13.2 实例解析

【实例要求】

熟悉 JDBC 数据库连接的基本环境、基本步骤和基本方法，同时能够掌握运用 JDBC 进行数据库的基本操作，包括查询、更新和调用存储过程等。

【实例解析】

1）JDBC 基本环境的建立

首先从网络上 http://www.microsoft.com/china/sql/downloads/2000/jdbc.asp 下载 SQL Server 2000 JDBC Driver。

2）JDBC 与 SQL Server 的连接应用

使用 JDBC-ODBC 桥创建与 SQL Server 数据库连接的步骤如下。

（1）SQL Server 中创建数据库 student 与数据表 stud，具体 SQL 语句如下：

```
CREATE DATABASE student;
CREATE TABLE stud
(sno char(8) PRIMARY KEY,
 sname char(10) NOT NULL,
 sage int NULL
);
```

(2) 建立数据库 student 的 ODBC 数据源,步骤如下:

① 打开 ODBC 数据源设置窗口,单击 User DSN 选项卡,单击 Add 按钮,选择 SQL Server 的 ODBC 驱动程序(只要安装 Microsoft SQL Server,便有 ODBC 驱动程序供该用户选择)。

② 单击"完成"按钮,开始 SQL Server 数据库的设置步骤。Name:指数据库在 ODBC 的名字,Java 在程序中引用该名字。Server:选择数据库所在的服务器,(local)代表本地服务器。

③ 单击"下一步"按钮后,设置访问 SQL Server 的用户名与密码。

④ 单击"下一步"按钮后,进入选择数据库的窗口,为数据源选择 student 数据库。

⑤ 单击"下一步"按钮后,进入设置数据库日程文件的保存路径。

⑥ 单击"完成"按钮完成数据库的设置。

(3) 创建 Java 代码连接 SQL Server 数据库,具体代码如下:

```java
package gdlgxy.shiyan;
    /* 功能描述:JDBC 与 SQL Server 的连接应用,建立数据库 student 的 ODBC 数据源,创
       建 Java 代码连接 SQL Server 数据库 */
import java.sql.*;
import java.io.*;
public class SqlServerDemo{
  public static void main(String[]args)throws Exception{
      //stud 指在 ODBC 上声明的数据库名称
      String url="jdbc:odbc:stud";
      String sqlstr,no,name;
      int age;
      Connection con;                          //建立连接类
      Statement stmt;                          //建立 SQL 语句执行类
      ResultSet rs=null;                       //建立结果集类
      //告诉程序使用 JDBC-ODBC 桥建立数据库连接
      Class.forName("sun.jdbc.odbc.JdbcOdbcDriver");
      //使用 DriverManager 类的 getConnection()方法建立连接
      //第一个字符参数定义用户名,第二个字符参数定义密码
      con=DriverManager.getConnection(url, "sa","");
      stmt=con.createStatement();              //创建 SQL 语句执行类
      sqlstr="select count(*) from stud";      //获取数据的记录类
      //使用 executeQuery()方法执行 SQL 命令
      rs=stmt.executeQuery(sqlstr);
      rs.next();
      //移到数据库的下一条记录
      //如果记录为空,那么加入十条记录
      if(rs.getint(1)==0){
          for (int i=1;i<11;++i){
              String  sql="insert  into  stud  values('"+Integer.toString(i)+"
','姓名 "+ Integer.toString(i)+"',"+
              Integer.toString(i)+") ";
              //使用 executeUpdate()方法执行查询以外的 SQL 命令
```

```
                stmt.executeUpdate(sql);
        }
    }
    //获得数据库的所有记录
    sqlstr="select * from stud";
    rs=stmt.executeQuery(sqlstr);
    //使用next()方法遍历数据库的每条记录
    while(rs.next()){
        //使用getString()方法取得字段的内容
        no=rs.getString(1);
        name=rs.getString(2);
        age=rs.getint(3);
        System.out.print("学号="+no);
        System.out.print(",姓名="+name);
        System.out.println(",年龄="+Integer.toString(age));
    }
    rs.close();                    //关闭结果集
    stmt.close();                  //关闭SQL语句执行类
    con.close();                   //关闭数据库连接类
  }
}
```

程序执行结果如下:

```
学号=1,姓名=姓名1,年龄=1
学号=10,姓名=姓名10,年龄=10
学号=2,姓名=姓名2,年龄=2
学号=3,姓名=姓名3,年龄=3
学号=4,姓名=姓名4,年龄=4
学号=5,姓名=姓名5,年龄=5
学号=6,姓名=姓名6,年龄=6
学号=7,姓名=姓名7,年龄=7
学号=8,姓名=姓名8,年龄=8
学号=9,姓名=姓名9,年龄=9
```

3) 运用 JDBC 进行数据库查询

下述示例是查询 SQL Server 本身自带的示例数据库 pubs 中数据表 Titles 的部分内容,具体代码如下:

```
package gdlgxy.shiyan;
    /* 功能描述:运用JDBC对已建立的数据库进行查询 */
import java.sql.*;
public class QueryDemo{
   public static void main(String[] args)throws Exception{
      //pubs 指在ODBC上声明的数据库名字
      String url="jdbc:odbc:pubs";
      String sqlstr,title_id,title;
```

```java
        Connection con;                          //建立连接类
        Statement stmt;                          //建立 SQL 语句执行类
        ResultSet rs=null;                       //建立结果集类
        //告诉程序使用 JDBC-ODBC 桥创建数据库连接
        Class.forName("sun.jdbc.odbc.JdbcOdbcDriver");
        //第一个参数定义用户名,第二个参数定义密码
        con=DriverManager.getConnection(url, "sa","");
        stmt=con.createStatement();
        //创建 SQL 语句执行类
        //获得数据库的所有记录
        sqlstr="select title_id,title from titles";
        rs=stmt.executeQuery(sqlstr);
        //使用 next()方法遍历数据库的每条记录
        while(rs.next()){
            //使用 getString()方法取得字段的内容
            title_id=rs.getString(1);
            title=rs.getString(2);
            System.out.print("title_id="+title_id);
            System.out.println(",title="+title);
        }
        rs.close();                              //关闭结果集
        stmt.close();                            //关闭 SQL 语句执行类
        con.close();                             //关闭数据库连接类
    }
}
```

4) 运用 JDBC 中 PreparedStatement 类创建带参数的 SQL 语句

下面以 SQL Server 本身自带的示例数据库 pubs 中数据表 Titles 为例,说明 PreparedStatement 类在数据查询中的应用,具体代码如下:

```java
package gdlgxy.shiyan;
    /* 功能描述:运用 JDBC 中 PreparedStatement 类创建带参数的 SQL 语句 */
import java.sql.*;
public class PreparedStatementDemo{
    public static void main(String[] args) throws Exception{
        //pubs 指在 ODBC 上声明的数据库名字
        String url="jdbc:odbc:pubs";
        String title_id,title;
        Connection con;                          //建立连接类
        PreparedStatement pStmt;                 //建立含参数 SQL 语句执行类
        //告诉程序使用 JDBC-ODBC 桥创建数据库连接
        Class.forName("sun.jdbc.odbc.JdbcOdbcDriver");
        //第一个参数定义用户名,第二个参数定义密码
        con=DriverManager.getConnection(url, "sa","");
        //创建带两个参数的 SQL 语句执行类
         pStmt = con.prepareStatement("update Titles set title =? where title_id=?");
```

```
        String[] strs1={"水浒传","西游记","三国演义","红楼梦"};
        String[] strs2={"BU1032","BU1111","BU2075","BU7832"};
        //更新数据表
        for (int i=0;i<4;++i){
            pStmt.setString(1,strs1[i]);            //设置第一个参数
            pStmt.setString(2, strs2[i]);           //设置第二个参数
            //使用 executeUpdate()方法执行带参数的 SQL 命令
            pStmt.executeUpdate();
        }
        pStmt.close();                              //关闭 SQL 语句执行类
        con.close();                                //关闭数据库连接类
    }
}
```

5) JDBC 中使用数据库的存储过程操作 SQL 命令

以 SQL Server 的 pubs 数据库为例说明运用 JDBC 调用数据库的存储过程,具体代码如下:

```
package gdlgxy.shiyan;
    /* 功能描述:JDBC 中使用数据库的存储过程操作 SQL 命令 */
import java.sql.*;
public class StoredProcedureDemo{
    static String sqlstr,title_id,title;
    static Connection con;                          //建立连接类
    static Statement stmt;                          //建立 SQL 语句执行类
    static ResultSet rs=null;                       //建立结果集类
    public static void main(String[]args)throws Exception{
        //pubs 指在 ODBC 上声明的数据库名字
        String url="jdbc:odbc:pubs";
        String strProc;
        //告诉程序使用 JDBC-ODBC 桥创建数据库连接
        Class.forName("sun.jdbc.odbc.JdbcOdbcDriver");
        //第一个参数定义用户名,第二个参数定义密码
        con=DriverManager.getConnection(url, "sa","");
        //创建 SQL 语句执行类
        stmt=con.createStatement();
        try{
            //创建 Srored procedure 的字符串
            strProc="create procedure showtitles as select * from titles";
            //向数据库写入 stored procedure 的 SQL 语句
            stmt.executeUpdate(strProc);
        }
        catch(Exception e){
        }
        //通过 connection 类的 prepareCall ()方法应用 stored procedure
        CallableStatement cs=con.prepareCall("{call showtitles}");
        //执行 stored procedure 的 SQL 语句
        rs=cs.executeQuery();
```

```
            showrecord();                    //显示数据集的记录
            rs.close();                      //关闭数据集
            stmt.close();                    //关闭SQL语句执行类
            con.close();                     //关闭数据库连接类
    }
    //显示数据集记录的方法
    static void showrecord() throws Exception{
        while(rs.next()){
            title_id=rs.getString(1);
            title=rs.getString(2);
            System.out.print("书籍的代号="+title_id);
            System.out.println(",书籍的名称="+title);
        }
    }
}
```

13.3 上机实验

【实验目的】

- 理解 JDBC 的功能及体系结构。
- 掌握利用 JDBC 实现数据库的查询更新等操作的方法。
- 掌握基本的 Java 数据库程序设计。

【实验要求】

设计一个小程序可以实现简单的学生管理系统。程序包括三个类,除了主程序及窗体程序外,数据库连接的功能单独由一个类来完成。

(1) 能够根据学生姓名、学号、班级、课程名称查询具体内容。
(2) 能够实现按照单科成绩、总成绩、平均成绩、学号排序。
(3) 能够实现学生信息的插入、删除和修改。
(4) 能够查询每门课程的最高分、最低分及相应学生姓名、班级和学号。
(5) 能够查询每个班级某门课程的优秀率(90 分及以上)、不及格率,并进行排序。
(6) 能够使用图形界面进行操作。

程序主界面及相关操作如图 13-1～图 13-4 所示。

图 13-1　程序主界面

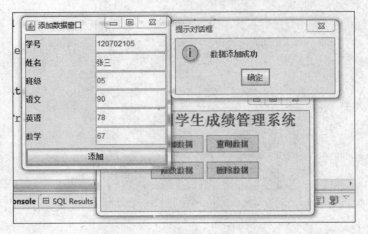

图 13-2　添加数据

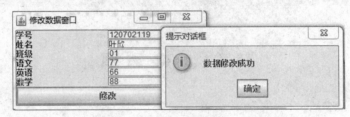

图 13-3　修改数据

图 13-4　删除数据

【实验指导】

本实验用到 java.util 包里的 Scanner 类的方法输入数据、java.lang 包里的 System.out.println()方法输出数据,输出内容要使用字符串拼接方法。

【程序模板】

连接数据库,代码如下:

```
package gdlgxy.shiyan;
    /* 功能描述:连接数据库,创建相应的数据库表 */
import java.sql.Connection;
import java.sql.DriverManager;
public class ConnectionUtil {
```

```
    public static Connection getConnection() throws Exception{
        Class.forName("com.mysql.jdbc.Driver");
        String url="jdbc:mysql://localhost:3306/JSD1407";
        String user="root";
        String password="1234";
        Connection conn=DriverManager.getConnection(url,user,password);
        return conn;
    }
    public static void main(String[] args) throws Exception{
        System.out.println(getConnection());
    }   //通过打印看一下是否连接上了
}       //将获得连接的方法封装
create table student(
no varchar(10) not null,
name varchar(50) not null,
class_no varchar(2) not null,
chinese double(3,1) not null,
math double(3,1) not null,
english double(3,1) not null,
primary key(no));
desc student;
insert into student(no,name,class_no,chinese,math,english) values(120702101,
'陈丽芳',1,66,77,88);
select * from student;
```

(1) MainFrame 类

```
package gdlgxy.shiyan;
    /* 功能描述:MainFrame 类 */
import javax.swing.*;
import java.awt.*;
import java.awt.event.*;
public class MainFrame extends JFrame{
    JButton insert,query,delete,modify;
    JPanel panel,panel1,panel2;
    public MainFrame()
    {
        //TODO Auto-generated method stub
        ImageIcon img=new ImageIcon("1.gif");
        JLabel text1,text2,picture=new JLabel(img);
        JFrame frame=new JFrame("学生成绩管理系统");
        insert=new JButton("添加数据");
        insert.setBackground(Color.green);
        insert.addActionListener(new insertActionPerformed());
        query=new JButton("查询数据");
        query.addActionListener(new queryActionPerformed());
        query.setBackground(Color.green);
        modify=new JButton("修改数据");
```

```java
            modify.setBackground(Color.green);
            modify.addActionListener(new modifyActionPerformed());
            delete=new JButton("删除数据");
            delete.setBackground(Color.green);
            delete.addActionListener(new deleteActionPerformed());
            frame.setSize(360,200);
            frame.setDefaultCloseOperation(JFrame.EXIT_ON_CLOSE);
            Container contentPane=frame.getContentPane();
            contentPane.setLayout(new BorderLayout());
            text1=new JLabel("欢迎使用学生成绩管理系统",JLabel.CENTER);
            text1.setFont(new Font("宋体",Font.BOLD,24));
            text1.setForeground(Color.blue);
            text2=new JLabel("------软工1班 yangz 制作");
            text2.setFont(new Font("TimesRoman",Font.ROMAN_BASELINE,14));
            panel1=new JPanel();
            panel1.add(insert);
            panel1.add(query);
            panel2=new JPanel();
            panel2.add(modify);
            panel2.add(delete);
            panel1.setOpaque(false);
            panel2.setOpaque(false);
            panel=new JPanel();
            panel.add(text2,BorderLayout.NORTH);
            panel.add(panel1,BorderLayout.NORTH);
            panel.add(panel2,BorderLayout.SOUTH);
            panel.setOpaque(false);
            contentPane.add(text1,BorderLayout.NORTH);
            contentPane.add(panel,BorderLayout.CENTER);
            frame.getLayeredPane().add(picture,new Integer(Integer.MIN_VALUE));
            Toolkit kit =Toolkit.getDefaultToolkit();
            Dimension screenSize =kit.getScreenSize();
            int screenWidth =screenSize.width/2;
            int screenHeight =screenSize.height/2;
            int height =this.getHeight();
            int width =this.getWidth();
            picture.setBounds(0,0,360,360);
            ((JPanel)contentPane).setOpaque(false);
            frame.setLocation(screenWidth-width/2, screenHeight-height/2);
            frame.setVisible(true);
}
public class insertActionPerformed implements ActionListener
{
    public void actionPerformed(ActionEvent e)
    {
        new Insert().setVisible(true);
    }
}
public class modifyActionPerformed implements ActionListener
```

```java
        {
            public void actionPerformed(ActionEvent e)
            {
                new Modify().setVisible(true);
            }
        }
        public class queryActionPerformed implements ActionListener
        {
            public void actionPerformed(ActionEvent e)
            {
                new Query().setVisible(true);
            }
        }
        public class deleteActionPerformed implements ActionListener
        {
            public void actionPerformed(ActionEvent e)
            {
                new Delete().setVisible(true);
            }
        }
        public static void main(String[] args)
        {
            new MainFrame();
        }
}
```

（2）Insert 类

```java
package gdlgxy.shiyan;
    /* 功能描述:Insert 类 */
import java.awt.*;
import java.awt.event.*;
import javax.swing.*;
import java.sql.*;
public class Insert extends JFrame
{
    JTextField input1,input2,input3,input4,input5,input6;
    JLabel label1,label2,label3,label4,label5;
    JButton button;
    static Statement st;
    static{
        try{
            Class.forName("com.mysql.jdbc.Driver");
            Connection con = DriverManager.getConnection(" jdbc: mysql://localhost:3306/jsd1407", "root", "123456");
            st=con.createStatement();
        }
        catch(Exception e){}
```

```java
    }
    ResultSet rs;
    public Insert()
    {
        input1=new JTextField(15);
        input2=new JTextField(15);
        input3=new JTextField(15);
        input4=new JTextField(15);
        input5=new JTextField(15);
        input6=new JTextField(15);
        JPanel panel=new JPanel();
        panel.setLayout(new GridLayout(6,2));
        panel.add(new JLabel("学号"),BorderLayout.CENTER);
        panel.add(input1);
        panel.add(new JLabel("姓名"),BorderLayout.CENTER);
        panel.add(input2);
        panel.add(new JLabel("班级"),BorderLayout.CENTER);
        panel.add(input3);
        panel.add(new JLabel("语文"),BorderLayout.CENTER);
        panel.add(input4);
        panel.add(new JLabel("英语"));
        panel.add(input5);
        panel.add(new JLabel("数学"));
        panel.add(input6);
        button=new JButton("添加");
        button.addActionListener(new mysql());
        Container container=getContentPane();
        container.add(panel,BorderLayout.CENTER);
        container.add(button,BorderLayout.SOUTH);
        setTitle("添加数据窗口");
        setDefaultCloseOperation(JFrame.DISPOSE_ON_CLOSE);
        setSize(250,250);
          Toolkit kit =Toolkit.getDefaultToolkit();
          Dimension screenSize =kit.getScreenSize();
          int screenWidth =screenSize.width/2;
          int screenHeight =screenSize.height/2;
          int height =this.getHeight();
          int width =this.getWidth();
          setLocation(screenWidth-width/2, screenHeight-height/2);
        setVisible(true);
    }
    class mysql implements ActionListener
    {
        public void actionPerformed(ActionEvent e)
        {
            try
            {
                String number=input1.getText().trim();
                String name=input2.getText().trim();
```

```
                    String class=input3.getText().trim();
                    String temp=input4.getText();
                    double chinese=Integer.parseInt(temp);
                    temp=input4.getText();
                    double english=Integer.parseInt(temp);
                    temp=input4.getText();
                    double maths=Integer.parseInt(temp);
                    if(number.equals("")|name.equals("")|class.equals("")|temp.equals(""))
                    {
                        JOptionPane.showMessageDialog(Insert.this,"请重新输入","提示对话框",1);
                    }
                    else
                    {
                        String sql="insert into Student(no,name,class_no,chinese,english,math) values('"+number+"','"+name+"','"+class+"',"+chinese+","+english+","+maths+");";
                        st.executeUpdate(sql);
                        JOptionPane.showMessageDialog(Insert.this,"数据添加成功","提示对话框",1);
                        input1.setText("");
                        input2.setText("");
                        input3.setText("");
                        input4.setText("");
                        input5.setText("");
                        input6.setText("");
                    }
                }
                catch(Exception ee){}
            }
        }
    }
```

（3）Modify 类

```
package gdlgxy.shiyan;
    /* 功能描述:Modify类 */
import java.awt.*;
import java.awt.event.*;
import javax.swing.*;
import java.sql.*;
public class Modify extends JFrame{
    JTextField input1,input2,input3,input4,input5,input6;
    JLabel label1,label2,label3,label4,label5;
    JButton button;
    static Statement st;
    static{
```

```java
        try{
            Class.forName("com.mysql.jdbc.Driver");
            Connection con = DriverManager.getConnection(" jdbc:mysql://localhost:3306/jsd1407","root","123456");
            st=con.createStatement();
        }
        catch(Exception e){}
    }
    ResultSet rs;
    public Modify()
    {
        input1=new JTextField(15);
        input2=new JTextField(15);
        input3=new JTextField(15);
        input4=new JTextField(15);
        input5=new JTextField(15);
        input6=new JTextField(15);
        JPanel panel=new JPanel();
        panel.setLayout(new GridLayout(6,2));
        panel.add(new JLabel("学号"));
        panel.add(input1);
        panel.add(new JLabel("姓名"));
        panel.add(input2);
        panel.add(new JLabel("班级"));
        panel.add(input3);
        panel.add(new JLabel("语文"));
        panel.add(input4);
        panel.add(new JLabel("英语"));
        panel.add(input5);
        panel.add(new JLabel("数学"));
        panel.add(input6);
        button=new JButton("修改");
        button.addActionListener(new mysql());
        Container container=getContentPane();
        container.add(panel,BorderLayout.CENTER);
        container.add(button,BorderLayout.SOUTH);
        setTitle("修改数据窗口");
        setDefaultCloseOperation(JFrame.DISPOSE_ON_CLOSE);
        setSize(300,150);
        Toolkit kit =Toolkit.getDefaultToolkit();
        Dimension screenSize =kit.getScreenSize();
        int screenWidth =screenSize.width/2;
        int screenHeight =screenSize.height/2;
        int height =this.getHeight();
        int width =this.getWidth();
        setLocation(screenWidth-width/2, screenHeight-height/2);
        setVisible(true);
    }
```

```java
        class mysql implements ActionListener
        {
            public void actionPerformed(ActionEvent e)
            {
                try
                {
                    String number=input1.getText().trim();
                    String name=input2.getText().trim();
                    String class=input3.getText().trim();
                    String temp=input4.getText();
                    double chinese=Integer.parseInt(temp);
                    temp=input4.getText();
                    double english=Integer.parseInt(temp);
                    temp=input4.getText();
                    double maths=Integer.parseInt(temp);
                    if(number.equals(""))
                    {
                        JOptionPane.showMessageDialog(Modify.this,"学号不能为空!","提示对话框",1);
                    }
                    else
                    {
                        try{
                        String sql="update Student set name='"+name+"',class_no='"+class+"',chinese="+chinese+",english="+english+",math="+maths+" where no='"+number+"'";
                        st.executeUpdate(sql);
                        JOptionPane.showMessageDialog(Modify.this, "数据修改成功","提示对话框",1);
                            input1.setText("");
                            input2.setText("");
                            input3.setText("");
                            input4.setText("");
                            input5.setText("");
                            input6.setText("");
                        }
                        catch(Exception ee)
                        {
                            JOptionPane.showMessageDialog(Modify.this,"请确认需要修改的学号是否存在","提示对话框",1);
                            System.out.println(ee);
                        }
                    }
                }
                catch(Exception eee)
                {
                    System.out.println(eee);
                }
```

            }
        }
}

(4) Query 类

```java
package gdlgxy.shiyan;
        /* 功能描述:Query 类 */
import java.awt.*;
import javax.swing.event.*;
import java.awt.event.*;
import javax.swing.*;
import java.sql.*;
public class Query extends JFrame{
    JTextArea show;
    JButton button1,button2,button3,button4,button5;
    JTextField field1,field2,field3;
    JComboBox comoBox;
    static Statement st;
    static{
        try{
            Class.forName("com.mysql.jdbc.Driver");
            Connection con = DriverManager. getConnection ( " jdbc: mysql://localhost:3306/jsd1407", "root", "123456");
            st=con.createStatement();
        }
        catch(Exception e){}
    }
    public Query()
    {
        show=new JTextArea(5,10);
        button1=new JButton("显示所有信息");
        button1.addActionListener(new Mysql1());
        Container container=getContentPane();
        container.setLayout(new BorderLayout());
        JPanel panel=new JPanel();
        JPanel mainpanel=new JPanel();
        button2=new JButton("按学号查询");
        button2.addActionListener(new Mysql2());
        panel.add(button2);
        field1=new JTextField(7);
        panel.add(field1);
        panel.setVisible(true);
        mainpanel.add(panel);
        button3=new JButton("按姓名查询");
        button3.addActionListener(new Mysql3());
        panel.add(button3);
        field2=new JTextField(6);
        panel.add(field2);
        panel.setVisible(true);
```

```java
            mainpanel.add(panel);
            button4=new JButton("按班级查询");
            button4.addActionListener(new Mysql4());
            panel.add(button4);
            field3=new JTextField(6);
            panel.add(field3);
            panel.setVisible(true);
            mainpanel.add(panel);
            String items[]={"请选择","语文","英语","数学"};
            comoBox=new JComboBox(items);
            button5=new JButton("按课程名称查询");
            button5.addActionListener(new Mysql5());
            panel.add(button5);
            panel.add(comoBox);
            panel.setVisible(true);
            mainpanel.add(panel);
            panel=new JPanel();
            panel.add(button1);
            container.add(mainpanel,BorderLayout.NORTH);
            container.add(panel,BorderLayout.SOUTH);
            container.add(new JScrollPane(show),BorderLayout.CENTER);
            setTitle("查询数据");
            setDefaultCloseOperation(JFrame.DISPOSE_ON_CLOSE);
            setSize(750,400);
              Toolkit kit =Toolkit.getDefaultToolkit();
              Dimension screenSize =kit.getScreenSize();
              int screenWidth =screenSize.width/2;
              int screenHeight =screenSize.height/2;
              int height =this.getHeight();
              int width =this.getWidth();
              setLocation(screenWidth-width/2, screenHeight-height/2);
            setVisible(true);
    }
    class Mysql1 implements ActionListener
    {
        public void actionPerformed(ActionEvent e)
        {
            try
            {
                String sql="select * from Student";
                ResultSet rs=st.executeQuery(sql);
                show.setText("");
                show.append("学号 姓名 班级 语文 英语 数学"+"\n");
                while(rs.next())
                {
                    show.append(rs.getString(1)+"          ");
                    show.append(rs.getString(2)+"          ");
                    show.append(rs.getString(3)+"          ");
                    show.append(rs.getDouble(4)+"          ");
                    show.append(rs.getDouble(5)+"          ");
```

```java
                    show.append(rs.getDouble(6)+"\n");
                }
            }
            catch(Exception ee){}
        }
    }
    class Mysql2 implements ActionListener
    {
        public void actionPerformed(ActionEvent e)
        {
            try
            {
                String ss=field1.getText().trim();
                String sql="select * from Student where no='"+ss+"'";
                ResultSet rs=st.executeQuery(sql);
                show.setText("");
                show.append("学号 姓名 班级 语文 英语 数学"+"\n");
                while(rs.next())
                {
                    show.append(rs.getString(1)+"          ");
                    show.append(rs.getString(2)+"          ");
                    show.append(rs.getString(3)+"          ");
                    show.append(rs.getDouble(4)+"          ");
                    show.append(rs.getDouble(5)+"          ");
                    show.append(rs.getDouble(6)+"\n");
                }
            }
            catch(Exception ee){}
        }
    }
    class Mysql3 implements ActionListener
    {
        public void actionPerformed(ActionEvent e)
        {
            try
            {
                String ss=field2.getText().trim();
                String sql="select * from Student where name='"+ss+"'";
                ResultSet rs=st.executeQuery(sql);
                show.setText("");
                show.append("学号 姓名 班级 语文 英语 数学"+"\n");
                while(rs.next())
                {
                    show.append(rs.getString(1)+"          ");
                    show.append(rs.getString(2)+"          ");
                    show.append(rs.getString(3)+"          ");
                    show.append(rs.getDouble(4)+"          ");
                    show.append(rs.getDouble(5)+"          ");
                    show.append(rs.getDouble(6)+"\n");
```

```java
            }
        }
        catch(Exception ee){}
    }
}
class Mysql4 implements ActionListener
{
    public void actionPerformed(ActionEvent e)
    {
        try
        {
            String ss=field3.getText().trim();
            String sql="select * from Student where class_no='"+ss+"'";
            ResultSet rs=st.executeQuery(sql);
            show.setText("");
            show.append("学号 姓名 班级 语文 英语 数学"+"\n");
            while(rs.next())
            {
                show.append(rs.getString(1)+"           ");
                show.append(rs.getString(2)+"          ");
                show.append(rs.getString(3)+"          ");
                show.append(rs.getDouble(4)+"          ");
                show.append(rs.getDouble(5)+"          ");
                show.append(rs.getDouble(6)+"\n");
            }
        }
        catch(Exception ee){}
    }
}
class Mysql5 implements ActionListener
{
    public void actionPerformed(ActionEvent e)
    {
        try
        {
            String sql="";
            String ss=comoBox.getSelectedItem().toString();
            if(ss.equals("语文"))
            {
              sql="select no,name,class_no,chinese from Student";
              show.setText("");
              show.append("学号 姓名 班级 语文"+"\n");
            }
            else if(ss.equals("英语"))
            {
              sql="select no,name,class_no,english from Student";
              show.setText("");
              show.append("学号 姓名 班级 英语"+"\n");
            }
```

```java
            else if(ss.equals("数学"))
            {
              sql="select no,name,class_no,math from Student";
              show.setText("");
              show.append("学号 姓名 班级 数学"+"\n");
            }
            ResultSet rs=st.executeQuery(sql);

            while(rs.next())
            {
                show.append(rs.getString(1)+"          ");
                show.append(rs.getString(2)+"          ");
                show.append(rs.getString(3)+"          ");
                show.append(rs.getDouble(4)+"\n");
            }
          }
          catch(Exception ee){}
      }
   }
}
```

(5) Delete 类

```java
package gdlgxy.shiyan;
    /* 功能描述:Delete 类 */
import java.awt.*;
import java.awt.event.*;
import javax.swing.*;
import java.sql.*;
public class Delete extends JFrame
{
      JButton search;
      JTextField input;
      JTextArea show;
      Connection con;
      Statement st;
    public Delete()
    {
      JPanel p=new JPanel();
      search=new JButton("删除");
      input=new JTextField(10);
      show=new JTextArea(6,43);
      p.add(new JLabel("输入要删除的学号"));
      p.add(input);
      p.add(search);
      setTitle("删除数据窗口");
      search.addActionListener(new Mysql());
      show.setEditable(false);
      add(p,BorderLayout.NORTH);
```

```java
        add(show,BorderLayout.CENTER);
        setSize(400,150);
        Toolkit kit =Toolkit.getDefaultToolkit();
        Dimension screenSize =kit.getScreenSize();
        int screenWidth =screenSize.width/2;
        int screenHeight =screenSize.height/2;
        int height =this.getHeight();
        int width =this.getWidth();
        setLocation(screenWidth-width/2, screenHeight-height/2);
        setVisible(true);
        validate();
        addWindowListener(new WindowAdapter()
        {
            public void windowClosing(WindowEvent e)
            {
                dispose();
            }
        });
    }

    class Mysql implements ActionListener
    {
        public void actionPerformed(ActionEvent ee)
        {
            try
            {
                Class.forName("com.mysql.jdbc.Driver");
            }
            catch(ClassNotFoundException eee)
            {
                System.out.println(""+eee);
            }
            try
            {
            String ss=input.getText().trim();
                con=DriverManager.getConnection("jdbc:mysql://localhost:3306/jsd1407", "root", "123456");
                st=con.createStatement();
                String sql="select * from Student where no='"+ss+"'";
                show.setText(sql);
                ResultSet rs=st.executeQuery(sql);
                while(rs.next())
                {
                    String  number=rs.getString("no");
                    String  name=rs.getString("name");
                    String  class=rs.getString("class_no");
                    double  chinese=rs.getInt("chinese");
                    double  english=rs.getInt("english");
                    double  math=rs.getInt("math");
```

```
                    show.setText("你删除了:\n");
                    show.append("学号:"+number+"\n姓名:"+name+"\n班级:"+class+
"\n语文"+chinese+"\n英语 "+english+"\n数学 "+math);
                    show.append("\n");
                    PreparedStatement ps=null;
                    sql="delete from Student where no=?";
                    ps=con.prepareStatement(sql);
                    ps.setString(1, ss);
                    ps.executeUpdate();
                }
                con.close();
            }
            catch(SQLException e)
            {
                show.setText("无此条记录");
                System.out.println(e);
            }
        }
    }
}
```

13.4 拓展练习

【基础知识练习】

一、选择题

1. 下列语句是关闭数据连接时使用的语句是(　　)。

   A. Statement SQL 语句变量＝连接变量.createStatement()

   B. Connection 连接变量＝DriverManager.getConnection(数据库 URL,用户账号,用户密码)

   C. Class.forName(JDBC 驱动程序名)

   D. 连接变量.close()

2. 下面的描述正确的是(　　)。

   A. PreparedStatement 继承自 Statement

   B. Statement 继承自 PreparedStatement

   C. ResultSet 继承自 Statement

   D. CallableStatement 继承自 PreparedStatement

3. 下面的描述错误的是(　　)。

   A. Statement 的 executeQuery()方法会返回一个结果集

   B. Statement 的 executeUpdate()方法会返回是否更新成功的 boolean 值

   C. 使用 ResultSet 中的 getString()可以获得一个对应于数据库中 char 类型的值

   D. ResultSet 中的 next()方法会使结果集中的下一行成为当前行

4. JDBC 提供的接口 java.sql.Statement 的功能是(　　)。
  A. 用于处理驱动程序的调入　　 B. 与特定数据库建立连接
  C. 用于 SQL 语句的执行　　 D. 用于保存查询所得的结果
5. 下列方法不是 Statement 对象所具有的方法的是(　　)。
  A. executeQuery()　　 B. executeUpdate()
  C. execute()　　 D. createStatement()
6. 下列选项不是 getConnection() 方法的参数的是(　　)。
  A. 数据库用户名　　 B. 数据库的访问密码
  C. JDBC 驱动器的版本　　 D. 连接数据库的 URL
7. 下列不属于更新数据库操作步骤的一项是(　　)。
  A. 加载 JDBC 驱动程序　　 B. 定义连接的 URL
  C. 执行查询操作　　 D. 执行更新操作

二、填空题

1. 在 JDBC 中可以调用数据库的存储过程的接口是_____。
2. 如果数据库中某个字段为 numberic 型,那么可通过结果集中的_____方法获取。
3. 在 JDBC 中使用事务,想要回滚事务的方法是_____类中的 rollback() 方法。
4. Statement 接口中的 executeQuery(String sql) 方法返回的数据类型是_____类的对象。

【上机练习】

  编写一个图形界面应用程序,利用 JDBC 实现班级学生管理,在数据库中创建 student 和 class 表格,应用程序具有以下功能。
  (1) 数据插入功能。能增加班级:在某班增加学生。
  (2) 数据查询功能。在窗体中显示所有班级,选择某个班级将显示该班的所有学生。
  (3) 数据删除功能。能删除某个学生,如果删除班级,则要删除该班所有学生。
  基本思路:每个功能提供一个操作面板,通过选项卡实现功能的切换。班级的选择采用下拉列表,由于班级的数量动态变化,因此班级选择下拉列表项也动态变化,这是设计中的一个难点。由于要对数据库数据进行增、删、改、查操作,在创建 Statement 对象时要注意选择参数,支持记录集与数据库的同步。

第 14 章

网 络 编 程

14.1 知 识 提 炼

14.1.1 网络编程基本概念

网络编程的目的就是直接或间接地通过网络协议与其他计算机进行通信。目前使用最广泛的网络协议是 Internet 上使用的 TCP/IP。网络编程中有两个主要的问题，一个是如何准确地定位网络上一台或多台主机，另一个是找到主机后如何可靠高效地进行数据传输。在 TCP/IP 中，IP 层主要负责网络主机的定位、数据传输的路由，由 IP 地址可以唯一地确定 Internet 上的一台主机。而 TCP 层则提供面向应用的可靠的或非可靠的数据传输机制，这是网络编程的主要对象，一般不需要关心 IP 层是如何处理数据的。

通常一台主机上总是有很多个进程需要网络资源进行网络通信。网络通信的对象准确地讲不是主机，而应该是主机中运行的进程。这时候仅有主机名或 IP 地址来标识这么多个进程显然是不够的。端口号就是为了在一台主机上提供更多的网络资源而采取的一种手段，也是 TCP 层提供的一种机制。只有通过主机名或 IP 地址和端口号的组合才能唯一地确定网络通信中的对象——进程。

服务类型(service)有 http、telnet、ftp、smtp 等。

服务类型是在 TCP 层上面的应用层的概念。基于 TCP/IP 可以构建出各种复杂的应用，服务类型是那些已经被标准化了的应用，一般都是网络服务器(软件)。读者可以编写自己的基于网络的服务器，但都不能被称作标准的服务类型。

14.1.2 两类传输协议：TCP 和 UDP

TCP(transmission control protocol)是一种面向连接的保证可靠传输的协议。通过 TCP 传输，得到的是一个顺序的无差错的数据流。发送方和接收方成对的两个 Socket 之间必须建立连接，以便在 TCP 的基础上进行通信，当一个 Socket(通常都是 Server Socket)等待建立连接时，另一个 Socket 可以要求进行连接，一旦这两个 Socket 连接起来，它们就可以进行双向数据传输，双方都可以进行发送或接收操作。

UDP(user datagram protocol)是一种无连接的协议，每个数据报都是一个独立的信息，包括完整的源地址或目的地址，它在网络上以任何可能的路径传往目的地，因此能否到达目的地，到达目的地的时间以及内容的正确性都是不能被保证的。

下面对这两种协议做简单比较。

使用 UDP 时，每个数据报中都给出了完整的地址信息，因此不需要建立发送方和接收方的连接。对于 TCP，由于它是一个面向连接的协议，在 Socket 之间进行数据传输之前必然要建立连接，所以在 TCP 中多了一个连接建立的时间。

使用 UDP 传输数据时是有大小限制的，每个被传输的数据报必须限定在 64KB 之内。而 TCP 没有这方面的限制，一旦连接建立起来，双方的 Socket 就可以按统一的格式传输大量的数据。UDP 是一个不可靠的协议，发送方所发送的数据报并不一定以相同的次序到达接收方。TCP 是一个可靠的协议，它确保接收方完全正确地获取发送方所发送的全部数据。

总之，TCP 在网络通信上有极强的生命力，例如，远程连接（Telnet）和文件传输（FTP）都需要不定长度的数据被可靠地传输。相比之下 UDP 操作简单，而且仅需要较少的监护，因此通常用于局域网高可靠性的分散系统中的 Client/Server 应用程序。

14.1.3　URL 的组成与创建

URL（uniform resource locator）是一致资源定位器的简称，它表示 Internet 上某一资源的地址，URL 的组成为 protocol://resourceName。协议名（protocol）指明获取资源所使用的传输协议，如 http、ftp、gopher、file 等，资源名（resourceName）则应该是资源的完整地址，包括主机名、端口号、文件名或文件内部的一个引用。例如：

http://www.sun.com/　　协议名://主机名

http://home.netscape.com/home/welcome.html　　协议名://机器名＋文件名

http://www.gamelan.com:80/Gamelan/network.html#BOTTOM　　协议名://机器名＋端口号＋文件名＋内部引用

一个 URL 对象生成后，其属性是不能被改变的，但是可以通过类 URL 所提供的方法来获取这些属性。

public String getProtocol()。获取该 URL 的协议名。

public String getHost()。获取该 URL 的主机名。

public int getPort()。获取该 URL 的端口号，如果没有设置端口，返回−1。

public String getFile()。获取该 URL 的文件名。

public String getRef()。获取该 URL 在文件中的相对位置。

public String getQuery()。获取该 URL 的查询信息。

public String getPath()。获取该 URL 的路径。

public String getAuthority()。获取该 URL 的权限信息。

public String getUserInfo()。获取使用者的信息。

创建一个 URL：

(1) public URL(String spec)；通过一个表示 URL 地址的字符串可以构造一个 URL 对象。

```
URL urlBase=new URL("http://www.263.net/")
```

(2) public URL(URL context, String spec)；通过基 URL 和相对 URL 构造一个 URL 对象。

```
URL net263=new URL ("http://www.263.net/");
URL index263=new URL(net263, "index.html")
```

（3）public URL(String protocol，String host，String file)；

```
new URL("http", "www.gamelan.com", "/pages/Gamelan.net.html");
```

（4）public URL(String protocol，String host，int port，String file)；

```
URL gamelan=new URL("http", "www.gamelan.com", 80, "Pages/Gamelan.network.html");
```

14.1.4 InetAddress 类

Internet 通过 IP 地址或域名标识主机，而 InetAddress 对象封装了两者的信息。以下是 InetAddress 类中定义的几个常用方法。

（1）static InetAddress getLocalHost()：返回本地主机对应的 InetAddress 对象。

（2）String getHostAddress()：返回 InetAddress 对象的 IP 地址。

（3）String getHostName()：返回 InetAddress 对象的域名。

14.1.5 Socket 通信原理

（1）客户端和服务端分别用 Socket 和 ServerSocket 类实现连接。

（2）Socket 的工作过程是服务端首先执行等待连接，根据指定端口建立 ServerSocket 对象，通过该对象的 accept()方法监听客户连接，客户端创建 Socket 对象请求与服务端的特定端口进行连接，连接成功后，双方将建立一条 Socket 通道，利用 Socket 对象的 getInputStream()和 getOutputStream()方法可得到对 Socket 通道进行读/写操作的输入/输出流，通过流的读/写实现客户与服务器的通信。

（3）Java Socket 通信编程经常结合多线程技术，一个服务器可以和多个客户机建立连接，进行并发通信。

14.1.6 Applet 对 URL 访问

（1）URL 的组成包括协议和资源，资源由主机名、端口、文件三部分构成。

（2）常用获取 URL 属性的方法有 getProtocal()、getHost()、getPort()、getFile()。

（3）利用 URLConnection 可获得 URL 资源的内容、内容长度、最后修改日期等各类属性以及对应 URL 连接的输入/输出流。

（4）Applet 调用 showDocument()方法在指定窗口中显示 URL 内容。

14.2 实例解析

【实例要求】

分析测试基于 TCP 的 C/S 网络编程（单客户）。

【实例源代码】

新建源文件客户 ShiyanTalkClient.java，源代码如下。

(1) 客户端程序

```java
package gdlgxy.shiyan;
    /* 功能描述:客户端程序 */
 import java.io.*;
 import java.net.*;
 public class ShiyanTalkClient {
 public static void main(String args[]) { try{
 Socket socket=new Socket("127.0.0.1",4700);
 //向本机的 4700 端口发出客户请求
 BufferedReader sin=new BufferedReader(new InputStreamReader(System.in));
 //由系统标准输入设备构造 BufferedReader 对象
 PrintWriter os=new PrintWriter(socket.getOutputStream());
 //由 Socket 对象得到输出流,并构造 PrintWriter 对象
 BufferedReader is = new BufferedReader (new InputStreamReader (socket.getInputStream()));
 //由 Socket 对象得到输入流,并构造相应的 BufferedReader 对象
 String readline;
 readline=sin.readLine(); //从系统标准输入读入的一字符串
 while(!readline.equals("bye")){
 //若从标准输入读入的字符串为 "bye",则停止循环
 os.println(readline);
 //将从系统标准输入读入的字符串输出到
 Server os.flush();
 //刷新输出流,使 Server 马上收到该字符串
 System.out.println("Client:"+readline);
 //在系统标准输出上打印读入的字符串
 System.out.println("Server:"+is.readLine());
 //从 Server 读入一字符串,并打印到标准输出上
 readline=sin.readLine();
 //从系统标准输入读入一字符串
 }    //继续循环
 os.close();         //关闭 Socket 输出流
 is.close();         //关闭 Socket 输入流
 socket.close();     //关闭 Socket
 }catch(Exception e) {
    System.out.println("Error"+e); //出错,则打印出错信息
 }
 }
```

(2) 服务器端程序

```java
package gdlgxy.shiyan;
    /* 功能描述:构造服务器端程序 */
```

```java
import java.net.*;
import java.applet.Applet;
public class ShiyanTalkServer{
    public static void main(String args[]) {
        try{
            ServerSocket server=null;
            try{
                server=new ServerSocket(4700);
                //创建一个 ServerSocket 在端口 4700 监听客户请求
            }
            catch(Exception e) {
                System.out.println("can not listen to:"+e);
                //出错,打印出错信息
            }
            Socket socket=null;
            try{
                socket=server.accept();
                /* 使用 accept()阻塞等待客户请求,有客户请求到来,则产生一个
                    Socket 对象,并继续执行 */
            }
            catch(Exception e) {
                System.out.println("Error."+e);                //出错,打印出错信息
            }
            String line;
            BufferedReader is = new BufferedReader (new InputStreamReader(socket.getInputStream()));
            //由 Socket 对象得到输入流,并构造相应的 BufferedReader 对象
            PrintWriter os=newPrintWriter(socket.getOutputStream());
            //由 Socket 对象得到输出流,并构造 PrintWriter 对象
            BufferedReader sin = new BufferedReader (new InputStreamReader(System.in));
            //由系统标准输入设备构造 BufferedReader 对象
            System.out.println("Client:"+is.readLine());
            //在标准输出上打印从客户端读入的字符串
            line=sin.readLine();
            //从标准输入读入的一字符串
            while(!line.equals("bye")){
                //如果该字符串为 "bye",则停止循环
                os.println(line);
                //向客户端输出该字符串
                os.flush();
                //刷新输出流,使 Client 马上收到该字符串
                System.out.println("Server:"+line);
                //在系统标准输出上打印读入的字符串
                System.out.println("Client:"+is.readLine());
                //从 Client 读入一字符串,并打印到标准输出上
                line=sin.readLine();
                //从系统标准输入读入的一字符串
            }    //继续循环
```

```
                os.close();              //关闭 Socket 输出流
                is.close();              //关闭 Socket 输入流
                socket.close();          //关闭 Socket
                server.close();          //关闭 ServerSocket
            }
            catch(Exception e){
                System.out.println("Error:"+e);    //出错,打印出错信息
            }
        }
    }
```

【实例解析】

(1) Socket 工作过程。

对于一个功能齐全的 Socket,都要包含以下基本结构,其工作过程包含以下四个基本的步骤。

① 创建 Socket。
② 打开连接到 Socket 的输入/出流。
③ 按照一定的协议对 Socket 进行读/写操作。
④ 关闭 Socket。

第 3 步是程序员用来调用 Socket 和实现程序功能的关键步骤,其他三步在各种程序中基本相同。

以上四个步骤是针对 TCP 传输而言的,使用 UDP 进行传输时略有不同,在后面会有具体讲解。

(2) 创建 Socket。

java 在包 java.net 中提供了两个类 Socket 和 ServerSocket,分别用来表示双向连接的客户端和服务端。这是两个封装得非常好的类,使用很方便。其构造方法如下:

```
Socket(InetAddress address, int port);
Socket(InetAddress address, int port, boolean stream);
Socket(String host, int port);
Socket(String host, int port, boolean stream);
Socket(SocketImpl impl);
Socket(String host, int port, InetAddress localAddr, int localPort);
Socket(InetAddress address, int port, InetAddress localAddr, int localPort);
ServerSocket(int port);
ServerSocket(int port, int backlog);
ServerSocket(int port, int backlog, InetAddress bindAddr);
```

其中,address、host 和 port 分别是双向连接中另一方的 IP 地址、主机名和端口号,stream 指明 Socket 是流 Socket 还是数据报 Socket,localPort 表示本地主机的端口号,localAddr 和 bindAddr 是本地机器的地址(ServerSocket 的主机地址),impl 是 Socket 的父类,既可以用来创建 ServerSocket,又可以用来创建 Socket。count 则表示服务端所能支持的最大连接数。例如:

```
Socket client =new Socket("127.0.01.", 80);
ServerSocket server =new ServerSocket(80);
```

注意：在选择端口时，必须小心。每一个端口提供一种特定的服务，只有给出正确的端口，才能获得相应的服务。0~1023 的端口号为系统所保留，例如，http 服务的端口号为 80，Telnet 服务的端口号为 21，FTP 服务的端口号为 23，所以在选择端口号时，最好选择一个大于 1023 的数以防发生冲突。在创建 Socket 时如果发生错误，将产生 IOException，在程序中必须对其作出处理。所以在创建 Socket 或 ServerSocket 时必须捕获或抛出例外。

（3）客户端的 Socket。

下面是一个典型的创建客户端 Socket 的过程。

```
try{
    Socket socket=new Socket("127.0.0.1",4700);
    //127.0.0.1 是 TCP/IP 协议中默认的本机地址
}catch(IOException e){
    System.out.println("Error:"+e);
}
```

这是最简单的在客户端创建一个 Socket 的小程序段，也是使用 Socket 进行网络通信的第一步，程序相当简单，在这里不作过多解释。

（4）服务器端的 ServerSocket。

下面是一个典型的创建 Server 端 ServerSocket 的过程。

```
ServerSocket server=null;
try {
    server=new ServerSocket(4700);
    //创建一个 ServerSocket 在端口 4700 监听客户请求
}
catch(IOException e){
    System.out.println("can not listen to :"+e);
}
Socket socket=null;
try {
    socket=server.accept();
    /* accept()是一个阻塞方法,一旦有客户请求,它就会返回一个 Socket 对象用于同客户进行交互 */
}
catch(IOException e){
    System.out.println("Error:"+e);
}
```

以上程序是 Server 的典型工作模式，只不过在这里 Server 只能接收一个请求，接收完后 Server 就退出了。实际的应用中总是让它不停地循环接收，一旦有客户请求，Server 总是会创建一个服务线程服务新来的客户,而自己继续监听。客户方和服务方都建立了用于通信的 Socket,接下来就是由各个 Socket 分别打开各自的输入/输出流。

(5) 打开输入/输出流。

类 Socket 提供了方法 getInputStream() 和 getOutStream() 来得到对应的输入/输出流以进行读/写操作,这两个方法分别返回 InputStream 和 OutputSteam 类对象。为了便于读/写数据,可以在返回的输入/输出流对象上建立过滤流,如 DataInputStream、DataOutputStream 或 PrintStream 类对象,对于文本方式流对象,可以采用 InputStreamReader 和 OutputStreamWriter、PrintWirter 等处理。例如:

```
PrintStream os=new PrintStream(new BufferedOutputStream(socket.getOutputStream()));
DataInputStream is=new DataInputStream(socket.getInputStream());
PrintWriter out=new PrintWriter(socket.getOutStream(),true);
BufferedReader in = new ButfferedReader ( new InputStreamReader ( Socket.
getInputStream()));
```

输入/输出流是网络编程的实质性部分,具体如何构造所需要的过滤流,要根据需要而定,能否运用自如主要看读者对 Java 中输入/输出部分掌握如何。

(6) 关闭 Socket。

每一个 Socket 存在时,都将占用一定的资源,在 Socket 对象使用完毕时,要将其关闭。关闭 Socket 可以调用 Socket 的 Close() 方法。在关闭 Socket 之前,应将与 Socket 相关的所有的输入/输出流全部关闭,以释放所有的资源。而且要注意关闭的顺序,与 Socket 相关的所有的输入/输出应首先关闭,然后再关闭 Socket。

```
os.close();
is.close();
socket.close();
```

尽管 Java 有自动回收机制,网络资源最终是会被释放的。但是为了有效地利用资源,建议读者按照合理的顺序主动释放资源。

从上面的两段程序中,我们可以看到 Socket 四个步骤的使用过程。读者可以分别将 Socket 使用的四个步骤的对应程序段选择出来,这样便于读者对 Socket 的使用有进一步的了解。

读者可以在单机上试验该程序,最好是能在真正的网络环境下试验该程序,这样更容易分辨输出的内容和客户机、服务器的对应关系。同时也可以修改该程序,提供更为强大的功能。

14.3 上机实验

【实验目的】

- 理解网络通信的原理和常用通信技术的概念。
- 掌握基于 TCP 协议的套接字的网络编程方法。
- 掌握网络通信在实际应用开发程序中的应用。

【实验要求】

分析测试基于 TCP 的 C/S 网络编程(多客户)。

【实验指导】

前面提供的 Client/Server 程序只能实现 Server 和一个客户的对话。在实际应用中，往往是在服务器上运行一个永久的程序，它可以接收来自其他多个客户端的请求，提供相应的服务。为了实现在服务器方给多个客户提供服务的功能，需要对上面的程序进行改造，利用多线程实现多客户机制。服务器总是在指定的端口上监听是否有客户请求，一旦监听到客户请求，服务器就会启动一个专门的服务线程来响应该客户的请求，而服务器本身在启动完线程之后马上又进入监听状态，等待下一个客户的到来。

客户端的程序和上面程序是完全一样的，读者可以跳过不读，把主要精力集中在 Server 端的程序上。

【程序模板】

(1) 客户端程序：MultiTalkClient.java

```java
package gdlgxy.shiyan;
        /* 功能描述:构建客户端程序.*/
import java.io.*;
import java.net.*;
public class MultiTalkClient {
    public static void main(String args[]) {
        try{
            Socket socket=new Socket("127.0.0.1",4700);
            //向本机的 4700 端口发出客户请求
            BufferedReader sin = new BufferedReader (new InputStreamReader(System.in));
            //由系统标准输入设备构造 BufferedReader 对象
            PrintWriter os=new PrintWriter(socket.getOutputStream());
            //由 Socket 对象得到输出流,并构造 PrintWriter 对象
            BufferedReader is = new BufferedReader (new InputStreamReader(socket.getInputStream()));
            //由 Socket 对象得到输入流,并构造相应的 BufferedReader 对象
            String readline;
            readline=sin.readLine();
            //从系统标准输入读入的一字符串
            while(!readline.equals("bye")){
                //若从标准输入读入的字符串为"bye",则停止循环
                os.println(readline);
                //将从系统标准输入读入的字符串输出到 Server
                os.flush();
                //刷新输出流,使 Server 马上收到该字符串
                System.out.println("Client:"+readline);
```

```java
            //在系统标准输出上打印读入的字符串
            System.out.println("Server:"+is.readLine());
            //从 Server 读入一字符串,并打印到标准输出上
            readline=sin.readLine();
            //从系统标准输入读入的一字符串
        }   //继续循环
        os.close();             //关闭 Socket 输出流
        is.close();             //关闭 Socket 输入流
        socket.close();         //关闭 Socket
    }
    catch(Exception e) {
        System.out.println("Error"+e);     //出错,则打印出错信息
    }
  }
}
```

(2) 服务器端程序:MultiTalkServer.java

```java
package gdlgxy.shiyan;
    /* 功能描述:构建服务器端程序 */
import java.io.*;
import java.net.*;
public class MultiTalkClient {
    public static void main(String args[]) {
        try{
            Socket socket=new Socket("127.0.0.1",4700);
            //向本机的 4700 端口发出客户请求
            BufferedReader sin = new BufferedReader ( new InputStreamReader
(System.in));
            //由系统标准输入设备构造 BufferedReader 对象
            PrintWriter os=new PrintWriter(socket.getOutputStream());
            //由 Socket 对象得到输出流,并构造 PrintWriter 对象
            BufferedReader is=new BufferedReader(new InputStreamReader(socket.
getInputStream()));
            //由 Socket 对象得到输入流,并构造相应的 BufferedReader 对象
            String readline;
            readline=sin.readLine();    //从系统标准输入读入的一字符串
            while(!readline.equals("bye")){
                //若从标准输入读入的字符串为"bye",则停止循环
                os.println(readline);
                //将从系统标准输入读入的字符串输出到 Server
                os.flush();
                //刷新输出流,使 Server 马上收到该字符串
                System.out.println("Client:"+readline);
                //在系统标准输出上打印读入的字符串
                System.out.println("Server:"+is.readLine());
                //从 Server 读入一字符串,并打印到标准输出上
                readline=sin.readLine();
```

```
                    //从系统标准输入读入的一字符串
        }           //继续循环
        os.close();                          //关闭 Socket 输出流
        is.close();                          //关闭 Socket 输入流
        socket.close();                      //关闭 Socket
    }
    catch(Exception e) {
        System.out.println("Error"+e);       //出错,则打印出错信息
    }
}
```

(3) 多线程程序：ServerThread.java

```
package gdlgxy.shiyan;
    /* 功能描述:多线程程序 */
import java.io.*;
import java.net.*;
public class ServerThread extends Thread{
    Socket socket=null;                      //保存与本线程相关的 Socket 对象
    int clientnum;                           //保存本进程的客户计数
    public ServerThread(Socket socket,int num) { //构造函数
        this.socket=socket;                  //初始化 socket 变量
        clientnum=num+1;                     //初始化 clientnum 变量
    }
    public void run() {                      //线程主体
        try{
            String line;
            BufferedReader is=new BufferedReader(new InputStreamReader(socket.getInputStream()));
            //由 Socket 对象得到输入流,并构造相应的 BufferedReader 对象
            PrintWriter os=newPrintWriter(socket.getOutputStream());
            //由 Socket 对象得到输出流,并构造 PrintWriter 对象
            BufferedReader sin = new BufferedReader ( new InputStreamReader(System.in));
            //由系统标准输入设备构造 BufferedReader 对象
            System.out.println("Client:"+clientnum +is.readLine());
            //在标准输出上打印从客户端读入的字符串
            line=sin.readLine();             //从标准输入读入的一字符串
            while(!line.equals("bye")){
                //如果该字符串为"bye",则停止循环
                os.println(line);            //向客户端输出该字符串
                os.flush();                  //刷新输出流,使 Client 马上收到该字符串
                System.out.println("Server:"+line);
                //在系统标准输出上打印该字符串
                System.out.println("Client:"+clientnum +is.readLine());
                //从 Client 读入一字符串,并打印到标准输出上
                line=sin.readLine();         //从系统标准输入读入的一字符串
            }                                //继续循环
```

```
            os.close();              //关闭Socket输出流
            is.close();              //关闭Socket输入流
            socket.close();          //关闭Socket
            server.close();          //关闭ServerSocket
        }
        catch(Exception e){
            System.out.println("Error:"+e);    //出错时,打印出错信息
        }
    }
}
```

14.4 拓展练习

【基础知识练习】

一、选择题

1. 如果在关闭 Socket 时发生一个 I/O 错误,会抛出(　　)。
   A. IOException   B. UnknownHostException
   C. SocketException   D. MalformedURLException

2. 当找不到客户的服务器地址时会抛出(　　)。
   A. IOException   B. UnknownHostException
   C. SocketException   D. MalformedURLException

3. 如果 DatagramSocket 构造函数不能正确地创建一个 DatagramSocket,会抛出(　　)。
   A. IOException   B. UnknownHostException
   C. SocketException   D. MalformedURLException

4. (　　)类的对象中包含有 Internet 地址。
   A. Applet   B. DatagramSocket
   C. InetAddress   D. AppletContext

5. 下面(　　)类能够支持 TCP/IP 连接。
   A. InetAddress   B. Packet   C. Socket   D. AppletContext

6. (　　)对象管理基于流的连接。
   A. ServerSocket   B. Socket   C. Vector   D. DatagramSocket

二、填空题

1. URL 是 _____ 的简称,它表示 Internet/Intranet 上的资源位置。这些资源可以是一个文件、一个目录或一个对象。

2. 每个完整的 URL 由四部分组成：_____、_____、_____ 以及 _____。

3. 两个程序之间只有在 _____ 和 _____ 方面都达成一致时才能建立连接。

4. Socket 称为 _____,也有人称为"插座"。在两台计算机上运行的两个程序之间有一个双向通信的链接点,而这个双向链路的每一端就称为一个 Socket。

5. Java.net 中提供了两个类：_____ 和 _____,它们分别用于服务器端和客户端

的 Socket 通信。

6. URL 和 Socket 通信是一种面向 _____ 的流式套接字通信,采用的协议是 _____ 协议。

【上机练习】

通过建立一个 Socket 客户端和一个 ServerSocket 服务端进行实时数据交换。运行时先打开服务器端,然后再打开客户端。单击客户端的"连接服务器"按钮将与服务器建立连接;连接之后,双方在各自图形界面的文本框中填写要发送的信息;然后单击 send 按钮,就可以进行通信了。程序运行结果如图 14-1 所示。整个程序分为客户端和服务器端两大部分,分别保存为 Client.Java 和 Server.Java 两个文件。

图 14-1　聊天程序运行界面(客户端和服务器端)